Victor Meyer

Game Ranger's Diary

Activities and observations of a ranger in
the African wilderness

VDM Verlag Dr. Müller

Impressum/Imprint (nur für Deutschland/ only for Germany)
Bibliografische Information der Deutschen Nationalbibliothek: Die Deutsche Nationalbibliothek verzeichnet diese Publikation in der Deutschen Nationalbibliografie; detaillierte bibliografische Daten sind im Internet über http://dnb.d-nb.de abrufbar.
Alle in diesem Buch genannten Marken und Produktnamen unterliegen warenzeichen-, marken- oder patentrechtlichem Schutz bzw. sind Warenzeichen oder eingetragene Warenzeichen der jeweiligen Inhaber. Die Wiedergabe von Marken, Produktnamen, Gebrauchsnamen, Handelsnamen, Warenbezeichnungen u.s.w. in diesem Werk berechtigt auch ohne besondere Kennzeichnung nicht zu der Annahme, dass solche Namen im Sinne der Warenzeichen- und Markenschutzgesetzgebung als frei zu betrachten wären und daher von jedermann benutzt werden dürften.

Coverbild: www.ingimage.com

Verlag: VDM Verlag Dr. Müller GmbH & Co. KG
Dudweiler Landstr. 99, 66123 Saarbrücken, Deutschland
Telefon +49 681 9100-698, Telefax +49 681 9100-988
Email: info@vdm-verlag.de

Herstellung in Deutschland:
Schaltungsdienst Lange o.H.G., Berlin
Books on Demand GmbH, Norderstedt
Reha GmbH, Saarbrücken
Amazon Distribution GmbH, Leipzig
ISBN: 978-3-639-33364-0

Imprint (only for USA, GB)
Bibliographic information published by the Deutsche Nationalbibliothek: The Deutsche Nationalbibliothek lists this publication in the Deutsche Nationalbibliografie; detailed bibliographic data are available in the Internet at http://dnb.d-nb.de.
Any brand names and product names mentioned in this book are subject to trademark, brand or patent protection and are trademarks or registered trademarks of their respective holders. The use of brand names, product names, common names, trade names, product descriptions etc. even without a particular marking in this works is in no way to be construed to mean that such names may be regarded as unrestricted in respect of trademark and brand protection legislation and could thus be used by anyone.

Cover image: www.ingimage.com

Publisher: VDM Verlag Dr. Müller GmbH & Co. KG
Dudweiler Landstr. 99, 66123 Saarbrücken, Germany
Phone +49 681 9100-698, Fax +49 681 9100-988
Email: info@vdm-publishing.com

Printed in the U.S.A.
Printed in the U.K. by (see last page)
ISBN: 978-3-639-33364-0

Copyright © 2011 by the author and VDM Verlag Dr. Müller GmbH & Co. KG and licensors
All rights reserved. Saarbrücken 2011

TABLE OF CONTENTS

	FOREWORD	2
JANUARY:	Umfolozi Game Reserve	3
FEBRUARY:	Mpenjati Nature Reserve	9
MARCH:	Oribi Gorge Nature Reserve	15
APRIL:	Rhino Bomas (Umfolozi)	21
MAY:	Rhino Bomas (Umfolozi)	29
JUNE:	Rhino Bomas (Umfolozi)	35
JULY:	Monks Cowl Nature Reserve	41
AUGUST:	Monks Cowl Nature Reserve	47
SEPTEMBER:	Monks Cowl Nature Reserve	53
OCTOBER:	Sodwana Bay Marine Reserve	59
NOVEMBER:	Sodwana Bay Marine Reserve	67
DECEMBER:	Sodwana Bay Marine Reserve	75
	BIOGRAPHY	79

FOREWORD

This is my official diary for the rangework conducted for Natal Parks Board (now *Ezemvelo KZN Wildlife*) during the year of 1991. It made out the third year of a higher/tertiary education programme for the National Diploma in Nature Conservation at Technikon Pretoria (now Tshwane University of Technology).

Although much care has been taken to accurately report the day-to-day tasks of a game or park ranger in the diverse province of KwaZulu-Natal, South Africa, this work ought not to be taken as ideal policy for rangers or other wildlife professionals in similar situations.

What enticed me to join the dedicated staff of Natal Parks Board was that they achieved local, national and international acclaim for persistent conservation and ecotourism over the last 100 years. Natal Parks Board is best known for bringing back the white rhino from the brink of extinction. I was privileged to be part of the largest and most prestigious wildlife auction in Africa. I was also involved with private conservation initiatives and traditional communities. My responsibilities were extremely varied. In general, I monitored fauna and flora, recorded observations, and supervised labouring staff. In addition, the opportunity to interact with people was particularly fulfilling both in the case of reserve visitors and neighbouring communities.

Numerous stories of adventure could no doubt be told, though they were never reported in a formal day's account. They range from surprising a black rhino bathing in the mud to a tiring horse slipping on a mountain ledge...

As for many of the duties of a new ranger, the leadership of the officer-in-charge is essential. The incredible support of Tom Yule (Umfolozi), Derek Kunz (Mpenjati), Brett Butland (Trafalgar), Rod Potter (Oribi Gorge), Apie Strauss (Rhino Bomas), George Zaloumis (Monks Cowl), and Dirk Rossouw (Sodwana Bay) is hereby acknowledged. Hindsight-clarity of Sodwana place names was kindly supplied by Mark White of SBL Scuba Centre.

Official days off in lieu of regular weekends and public holidays were granted on the basis of five days per month. A further acknowledgement is due to David and Marié Avery for allowing me to stay over at their holiday apartment in Ballito for a significant amount of time.

Victor Meyer

JANUARY

1 January

Public Holiday.

2 January

Head Office administration (Queen Elizabeth Park, Pietermaritzburg).

3 January

To station.

4 January Mbhuzane Outpost, Umfolozi Game Reserve

We looked for two African buffalo outside the reserve near Engeni – no success. We also looked for a white rhino that broke out through the fence and found it running around in a populated, rural region. We tried to get it stabilized by cordoning off the area surrounding it. No success was to be had. It ran off in the direction of the Corridor (adjoining wilderness zone between Umfolozi and Hluhluwe Game Reserves) where it got trapped in a drainage channel. We waited for the head warden's arrival but he was delayed. Meanwhile, we opened a gate to the Corridor and tried to chase the rhinoceros towards the gate, but without success. It ran up along the fence and eventually broke through to the Corridor.

5 January Mbhuzane Outpost, Umfolozi Game Reserve

I went out with the patrol section to explore the immediate area. One black rhino was monitored and the data recorded. Identification was by means of ear notches and horn curvature. Cattle were chased out of the reserve from the banks of the Black Mfolozi River, and the gate was subsequently locked at 1800 hours.

6 January Mbhuzane Outpost, Umfolozi Game Reserve

I went out with the patrol section in the morning to investigate the area and to open up the Unkolotshe Gate. The idea of the gate was to make natural water accessible to neighbouring people and domestic animals because of a water shortage in the inhabited area. Water supply was as per the Black Mfolozi River. Cattle, however, may not graze on the grasslands inside the reserve, which inadvertently happens quite frequently. The gate was closed later in the evening when people and their cattle have left the reserve.

7 January   Mbhuzane Outpost, Umfolozi Game Reserve

I went out with the patrol section to open Unkolotshe Gate to the neighbouring community, while patrolling the surrounding area. The difference between black rhino tracks (round or flat back) and white rhino tracks (notched back) was observed. The gate was closed again after the people had gone back to their huts.

8 January   Mbhuzane Outpost, Umfolozi Game Reserve

I went out with the patrol section to patrol the area and to open and close the community gate. It's possible to decommission the gate by forfeiting roughly four hectares (ten acres) of reserve land to the adjacent Zulu tribe in order to extend full water rights to them. However, overutilization will turn the erosion-prone area into unusable, trampled land resulting in a polluted Black Mfolozi River. General repairs to the fence were carried out on patrol.

9 January   Mbhuzane Outpost, Umfolozi Game Reserve

I went out with the corporal to inspect certain game guard camps for cleanliness and to check the identification documents of occupants. Picnic sites were also visited to collect refuse.

10 January   Mbhuzane Outpost, Umfolozi Game Reserve

I went out on a large-scale patrol to Sokhwezela via Khukhu in order to investigate a case of poaching. A black rhino's horns were butchered out and the carcass left behind. The area was well known for such violations. In the afternoon, I went out on tractor with the gardeners to collect plants for supplementing the garden beds back at the outpost.

11 January   Mbhuzane Outpost, Umfolozi Game Reserve

I went out on patrol through Ngutshini to investigate the boundary fence. Signs of poachers were evident, as the fence was clipped. In the afternoon, we mended the fence by patching the gaps with wire mesh before locking Unkolotshe Gate.

12 January   Mbhuzane Outpost, Umfolozi Game Reserve

I went out with the patrol section to unlock Unkolotshe Gate and to conduct a general investigation of the area until the afternoon. A stray dog (problem animal) was destroyed for entering the restricted zone. The area was yet again cleared of people and cattle before locking the gate at sunset.

13 January   **Mbhuzane Outpost, Umfolozi Game Reserve**

I went out with the game guards to patrol the area. Unkolotshe Gate was also unlocked. The sighting of a black rhino was recorded for monitoring purposes. Later in the afternoon the gate was locked again.

14 January   **Mbhuzane Outpost, Umfolozi Game Reserve**

I went out on patrol and opened Unkolotshe Gate, then proceeded towards Ngutshini camp before securing the gate and returning to base in the afternoon.

15 January   **Mbhuzane Outpost, Umfolozi Game Reserve**

I went on patrol of the general area. The idea had been to spot signs of poached rhino or snared animals, tracks of poachers as well as other evidence of their suspected presence, such as fireplace remains. It must be stressed that time schedules and patrol routes are frequently changed in order to enlarge the surprise factor.

16 January   **Mbhuzane Outpost, Umfolozi Game Reserve**

I went out for shooting practice at the range with R1 rifles at distances of 50 metres (all stances) and 25 metres (rapid fire). My final score was 84 per cent. I embarked on a general patrol in the late afternoon, and reported no irregularities.

17 January   **Mbhuzane Outpost, Umfolozi Game Reserve**

I went on general patrol and discovered a male blue wildebeest calf which probably died of the cold, wet weather the previous evening. I went out on general patrol in the afternoon again; what was thought to have been would-be poachers turned out to be nearby residents.

18 January   **Mbhuzane Outpost, Umfolozi Game Reserve**

I went out on patrol with the primary objective of finding lion tracks outside the reserve. Apparently a lion had broken out and remained a threat to nearby residents and their cattle until it was located, darted/tranquilized and transferred to the inside of the reserve boundaries. Success was not to be claimed by us, but by another group who found the lion first.

19 January　　　　　　　　　　**Mbhuzane Outpost, Umfolozi Game Reserve**

We went out on general patrol – the idea had been for me as novel ranger to get familiarized with the section area. We stealthily waited in guard on Zintunzini Hill in the late afternoon because of armed men earlier seen in the area. It was thought that they might poach rhino. High-power binoculars and other equipment were used to scan the entire area outside the reserve's boundaries, but no sightings were made.

20 January　　　　　　　　　　**Mbhuzane Outpost, Umfolozi Game Reserve**

We went out to patrol the area at Umthumbhu which forms the boundary of Mbhuzane Section. It is adjacent to the southern part of the Corridor and Eastern Section. The first priority was to look for vantage points for the black rhino monitoring project. I went on a general patrol in the afternoon near Nquyini.

21 January　　　　　　　　　　**Madlozi Outpost, Umfolozi Game Reserve**

I was posted to the Madlozi outpost in order to get familiar with it as part of the reserve section in which I was working, having Mbhuzane as base. I went out to Lubisana Hill in the afternoon for general orientation and to spot any irregularities.

22 January　　　　　　　　　　**Madlozi Outpost, Umfolozi Game Reserve**

We went out for a general patrol which was interrupted by an emergency call to search out a problem lion that had been spotted outside the reserve in the direction of Engeni Gate nearby Zululand Anthracite Colliery (coalmine). Unfortunately, we couldn't find any trace of the lion. Coincidentally however, outlaws were caught red-handed with unlicensed pellet guns (air rifles) and some native birds that they had shot.

23 January　　　　　　　　　　**Madlozi Outpost, Umfolozi Game Reserve**

I went out on patrol of almost the entire region, with the priority of checking the fence line for dislodged droppers, etc. Some black rhino were also monitored. We went by the Mhlolokazana guard camp in the end.

24 January　　　　　　　　　　**Madlozi Outpost, Umfolozi Game Reserve**

I went out on patrol of the mountain range, consisting of Zintunzini, Msasaneni and Sabokwe, to check the entire fence line running along the boundary. We turned off at Engeni Gate. A previous poaching site was noticed as well as a lookout spot.

25 January　　　　　　　　　　**Madlozi Outpost, Umfolozi Game Reserve**

I went on patrol to Umphafa River (Nkawu) and passed nearby Mantiyane Hill to check the visitor hide. We patrolled the areas between Lubisana and Sabokwe, Msasaneni, Zintunzini and Sokhwezela. I was posted back to Mbhuzane in the late afternoon. It must be stressed that, during field excursions, diverse signs of the wild, such as animal tracks, droppings, sounds, and behaviour are frequently studied. Grass and tree samples are also collected for later identification by means of field guides and other books – same goes for the identification of birds, frogs, snakes and other reptiles, antelope and other ungulates, carnivores, insects, spiders, etc. The Zulu language was being learnt continually.

26 January　　　　　　　　　　**Mbhuzane Outpost, Umfolozi Game Reserve**

I went out on general patrol and observed the fence line from a viewpoint in order to monitor any intruders – no violations occurred. Back at base, I joined in a staff meeting for the game guards and labourers. I was given briefings on my new station for February.

27 January　　　　　　　　　　**Mbhuzane Outpost, Umfolozi Game Reserve**

Official Day Off (spent at Ballito, Dolphin/North Coast).

28 January　　　　　　　　　　**Mbhuzane Outpost, Umfolozi Game Reserve**

Official Day Off (spent at Ballito, Dolphin/North Coast).

29 January　　　　　　　　　　**Mbhuzane Outpost, Umfolozi Game Reserve**

Official Day Off (spent at Ballito, Dolphin/North Coast).

30 January　　　　　　　　　　**Mbhuzane Outpost, Umfolozi Game Reserve**

Official Day Off (spent at Ballito, Dolphin/North Coast).

31 January　　　　　　　　　　**Mbhuzane Outpost, Umfolozi Game Reserve**

Official Day Off (spent at Ballito, Dolphin/North Coast).

Date: 25/01/91   Station: UMFOLOZI / MADLOZI

Activities/Observations:

Went out on patrol to Mphafa River (Nkawu) and checked out the visitor hide. Patrolled area between Lubisana and Sabokwe, Msasaneni, 22 Intonzini, Sokwezele. Went nearby Mantiyane Hill as well.

Posted back to Mbuzane in the late afternoon.

← P.S.: It must be stressed that on field excursions diverse signs of the wild like spoor, dung, noise, behaviour etc. are pointed out frequently. Grass and tree samples are also collected for later identification by means of field guides and books - same goes for identification of birds, frogs, snakes and other reptiles, antelope and other species of animal, carnivores & insects (spiders) etc.

← The Zulu language is also being studied frequently.

FEBRUARY

1 February   **Mpila Rest Camp, Umfolozi Game Reserve**

We organized labour early in the morning. I sat in at a disciplinary hearing concerning two employees who were absent from work without leave – they received written warnings. We checked on the infrastructure development at Mdindini Camp and found it to be satisfactory.

2 February   **Mpenjati Nature Reserve, South Coast**

Arrival at new station: familiarizing and orientation of the basic setup that stands.

3 February   **Mpenjati Nature Reserve, South Coast**

I went out on patrol for detailed familiarization and orientation. Policy towards the reserve's neighbours dictated that agricultural rights still allowed harvesting on farms in the conservancy. Anti-erosion structures for retention of dunes were observed as well as boardwalks over soggy areas (wetland). The sand-winning operation site was also noticed. Developmental phases were discussed and the completed infrastructure appreciated. Basic plant and animal life (ecology) was considered as well as the practical implications of combating alien plants.

4 February   **Mpenjati Nature Reserve, South Coast**

I discussed initiatives of developing a new trail with the officer-in-charge. We walked and investigated the proposed area for trail layout. The standing infrastructure was examined and offensives were drawn up for future development under a scheduled plan. Filed documents and plans of previous years were looked through in order to get familiarized with their content and adding to them.

5 February   **Mpenjati Nature Reserve, South Coast**

I filled out the monthly visitor chart by interpreting and summarizing the raw data, calculating towards resultant totals. New tendencies could be noticed in comparison with previous months, reflecting the peak times of the year. Further construction to the existing footpath was done so as to connect with the layout of the new trail. Levelling of the surface, embedding of log mats, and verging with rocks on both sides of the path were the tasks at hand. Log mats were only laid down on erodible slopes. The clearing of vegetation for accessibility was minimized and aesthetic appeal borne in mind by binding up the overhung branches with blend-in rope.

6 February   **Mpenjati Nature Reserve, South Coast**

Roll-call of the labour force was taken first thing in the morning as done every day. We continued with construction work to the trail, especially the clearing of thick bush by use of a chainsaw. We started manufacturing more log mats to replenish the supply of surface cover.

7 February   **Mpenjati Nature Reserve, South Coast**

We carried on with the making of log mats. Later in the day, the manufacturing of garden benches started up for installation at points of rest and reflection on the trail.

8 February   **Mpenjati Nature Reserve, South Coast**

We had completed sufficient log mats for the time being, and continued with the manufacturing of benches. Furthermore, we packed confinement rocks on both sides of the trail path, filling up the verge edges, making slight alterations to direction and position of the trail as well as cleaning up the adjacent areas.

9 February   **Mpenjati Nature Reserve, South Coast**

We completed bench-making and continued with rock confinement (verging) of the trail. Placement of log mats over erodible and slippery slopes was also carried out on the path.

10 February   **Mpenjati Nature Reserve, South Coast**

We continued with rock confinement and lifting of verge edges, stabilizing the affected surface of the trail. Basic tidying of the first section (up to the footbridge) was done. Later in the day, a continuing weekly survey for the Oceanic Research Council was conducted. It involved the collection of sand and water samples, measurements of tide, crest, estimated depth at river mouth, fluctuation of estimated water level, specific gravity (for subsequent conversion to salinity), water temperature and, of course, wind speed and direction (by ventimeter). All measurements were standardized at set points of locality.

11 February   **Mpenjati Nature Reserve, South Coast**

We went further with the section of the trail on the other side of the footbridge in terms of rock confinement (verge-making), path levelling, the laying of log mats onto steep, erodible areas of the path, and some bush clearance.

12 February   **Mpenjati Nature Reserve, South Coast**

We continued with path confinement using rocks and log mat installation on the trail. A section of the route was altered to improve accessibility and at the same time avoid removal of an established tree – thus retaining the forest canopy. I partly succeeded in trying to find an alternative route for another section of the trail beyond a junction point. This was to minimize the steepness of the trail by following a contour line as far as possible.

13 February   **Mpenjati Nature Reserve, South Coast**

We installed the garden benches previously manufactured, after I had had a look for suitable sites for these further along the path in anticipation of the position of the trail that might be extending. The few possibilities that came up were first clarified.

14 February   **Mpenjati Nature Reserve, South Coast**

We went further along the trail, installing garden benches and tidying the surrounding areas. Overhanging branches were tied up so as to make the trail more accessible but still aesthetically pleasing to visitors.

15 February   **Mpenjati Nature Reserve, South Coast**

We continued with the binding up of obstructive branches and general tidying of adjacent areas on the trail. Replacement of log mats was done because of route alteration. In the afternoon, we were briefed at the nearby and southernmost Umtamvuna Nature Reserve on the objections of developing a trail for school groups, etc. The project was envisaged to start as from March that year.

16 February   **Umtamvuna Nature Reserve, South Coast**

As a delegation of rangers and wardens, we explored potential educational sites along an anticipated trail by vehicle because of rain. Catchment principles were discussed and rural applications in terms of tribal traditions/politics interpreted. Thus, the planting of crops and its negative effects on the wetland raised the question: Should the local people change their lifestyle or land practices? We visited Caribbean Estates to investigate their infrastructure and how lightly the development impacted upon the conservation area. The trails were considered to be excellent. We returned to Mpenjati later in the day.

17 February   **Mpenjati Nature Reserve, South Coast**

We continued with the laying down of log mats along the trail. Some mats were first assembled. The Oceanic Research Council survey was done like reported a week ago.

Turbidity and current/stream velocity were also ascertained. Later in the day, I went on a field excursion to explore and identify some endemic trees of the area.

18 February   Mpenjati Nature Reserve, South Coast

We went further with rock confinement (verging) of the trail and log mat manufacturing, among other duties. We also adjusted the suspension footbridge by tightening the necessary links (turnbuckles, D-shackles) in places.

19 February   Mpenjati Nature Reserve, South Coast

We continued with log mat manufacturing, rock confinement (verging) and laying down of mats on the trail. Unsatisfactory sections of the route were slightly altered. Excessive bush was cleared for access and aesthetic reasons by the binding up of branches, whereas smaller plants obstructing the path were transplanted to fit into the adjacent habitat.

20 February   Mpenjati Nature Reserve, South Coast

Official Day Off (spent at base).

21 February   Mpenjati Nature Reserve, South Coast

Official Day Off (spent at base).

22 February   Mpenjati Nature Reserve, South Coast

Official Day Off (spent at base).

23 February   Mpenjati Nature Reserve, South Coast

Official Day Off (spent at base).

24 February   Mpenjati Nature Reserve, South Coast

Official Day Off (spent at base).

25 February Trafalgar Marine Reserve, South Coast

I went out with the zone officer and game guards to do shopping in town because of payday. The marine conservancy was explored from Southbroom to Port Edward. Its policy on extensive catchment conservation, public and business relations as well as basic law enforcement were discussed.

26 February Trafalgar Marine Reserve, South Coast

We went to San Lameer to make new arrangements in terms of policy and goal-setting because of new management concerning the conservancy. We then travelled to Port Edward to make further arrangements with the ski-boat club regarding the sea-fishing competition and festival which takes place annually in the autumn. Competition rules were made according to nature conservation laws. We subsequently went to Southbroom to resolve a pollution problem together with a water affairs officer who had taken some water samples at various points of the river. Tests should be positive for timber treatment substances used at the construction site of a new log cabin development.

27 February Trafalgar Marine Reserve, South Coast

I went out with the Trafalgar game guards to patrol the area from Mpenjati to Marina Beach. No crimes of the intertidal zone were detected. The life forms in rock pools were noted and laws applicable to seafood collection by licensees or permit holders considered. Public rapport was built through conversation with beachgoers. At springtide, I took the game guards to Glenmore Beach by vehicle to stand watch through the evening for law enforcement purposes.

28 February Trafalgar Marine Reserve, South Coast

We investigated an occurrence of leguan lizard in the roof of a Southbroom residence but were unsuccessful in capturing it because of its agility. We also patrolled Marina Beach and surrounds by vehicle for general law enforcement. A reported case of an encaged squirrel monkey was investigated, but the owner was absent. We finished the day's duties by checking on the launch site for ski-boats at Port Edward.

Date: 19/02/91 Station: MPENJATI

Activities/Observations:

Went further with mat manufacturing, stone confinement, lodging, laying out of mats, alttering of poor path at unsatisfying sections, bush clearing for esthetically pleasingness, ~~tideing up~~ binding up of branches etc., transplanting of plants in the path to Mabiti & adjacent area.

signature

Duties carried out & as reflected in this diary for the period 2/02/1991 - 19/02/1991, are hereby verified.

signature
WARDEN MPENJATI
1991/02/19

MARCH

1 March   Oribi Gorge Nature Reserve, South Coast

I received basic orientation of the reserve as well as briefings on the tasks at hand. Some trail construction was done at Hoopoe Falls by means of rock confinement, verging, levelling and smoothing of the surface.

2 March   Oribi Gorge Nature Reserve, South Coast

I did the monthly financial report in terms of fuel expenditure, cash flow, budgeting for orders, maintenance costs and phone bills. I also accompanied a tourist group from the Banana Express on the Baboon Trail, listening to interpretive talks and extension on the basic ecology of the reserve and its natural wonders as well as principles and policies of the greater conservancy and its buffering effect on the reserve as a hub.

3 March   Oribi Gorge Nature Reserve, South Coast

I went out with the game guards on patrol to the picnic site via the View Trail for maintaining law and order and building rapport with members of the public.

4 March   Oribi Gorge Nature Reserve, South Coast

After the labourers had been dropped off to do bush clearing on the various trails, I prepared and primed a signboard for painting. I also fitted block-mounted photograph enlargements with screw eyes and picture-hanging wire to decorate the walls of the tourist rondavels (accommodation). I went to Paddock Post Office later in the day to collect mail, before picking up the labourers. At the same time, the picnic site's ablutions were checked upon.

5 March   Oribi Gorge Nature Reserve, South Coast

I went along with the officer in charge of Oribi Gorge to meet up with the officer at Mpenjati Reserve for a discussion on conservancy programmes and timeframes for the remaining year, upon which some ideas were highlighted for the future. The Conservancy Day in March was clearly defined as will the others be once they've drawn near. After returning from the meeting, I painted the signboard, prepared the previous day, in black enamel for the base and green for the sign.

6 March   **Oribi Gorge Nature Reserve, South Coast**

I gave the signboard a final coat of paint and fitted more framed prints with screw eyes and picture wire for the decoration of the cottage/squaredavel. I subsequently fixed the roof of the ranger's house by fitting new roofing screws and applying bitumen sealer to stop any future leakage. I later went out with a guided Banana Express group on the regular Baboon Trail.

7 March   **Oribi Gorge Nature Reserve, South Coast**

I attended the onsite annual management meeting to discuss a vast variety of suggestions, goals and accomplishments. For instance, the tourist development scheme, water provision, recreation (swimming pool and catering area), fire programme for the coming year and thereafter, anti-erosion measures on roads (surface upgrade), alien plant control programme, and preservation of unique habitats.

8 March   **Oribi Gorge Nature Reserve, South Coast**

I went out with a high school group from Durban to give interpretative education and guidance on the View and Baboon trails. Back at base, I repainted the chalkboards with a first coat after preparation by fine sanding was completed.

9 March   **Oribi Gorge Nature Reserve, South Coast**

I gave the blackboards around the rest camp a final coat of paint. I also painted the pickup truck's rails with a final coat of epoxy. Earlier in the day, I went with the visiting Banana Express group on the Baboon Trail to give interpretation to them.

10 March   **Oribi Gorge Nature Reserve, South Coast**

I applied the final touch-ups of paint to the pickup truck's rails and other places on the vehicle susceptible to rust, such as chips in the body resulting from road stones. I later joined the Paddock and Plains conservancy meeting to discuss the annual freshwater fishing day arrangements and additional twilight activities – bird and game viewing.

11 March   **Oribi Gorge Nature Reserve, South Coast**

I repainted a section of blackboard (script lines to follow). I also treated the moist walls of the ablutions interiorly with sealer and applied an enamel coating.

12 March   **Oribi Gorge Nature Reserve, South Coast**

I went out with the labour team to continue work on the Hoopoe Falls trail by means of stone packing, verging, and levelling its surface. Bush clearance for accessibility was to be kept as ecologically considerate and aesthetically pleasing as possible.

13 March   **Oribi Gorge Nature Reserve, South Coast**

I went out with semi-weekly Banana Express visitors on the Baboon Trail for interpretation of the bush environment. I examined the basic plant life by identifying common tree species. Invasive plants were noted. Insect life, birdlife, and mammalian life were frequently inspected.

14 March   **Oribi Gorge Nature Reserve, South Coast**

I took the labourers to the official houses of the conservator and another officer to do some gardening. The gardens of nearby staff compounds were also attended to. Back at base, I fixed the roof of a pump house (my temporary single-quarter accommodation) by replacing some roofing screws and sealing in places with bitumen.

15 March   **Oribi Gorge Nature Reserve, South Coast**

I went out with the Paddock Fishing Club to put up signage for the weekend's angling competition and to familiarize myself further with the conservancy area. Water pH was measured at various fishing spots along a series of dams (stocked with black bass) and the implications for fish behaviour were discussed.

16 March   **Oribi Gorge Nature Reserve, South Coast**

I went to the angling competition, organized by the Paddock and Plains Farmers' Association, to oversee and assist with the weighing of fish caught during the event. Basic law enforcement in terms of fish-size requirements as well as issuing and checking of fishing licenses was done. Improvement of public relations and extension work were carried out throughout the day. Fish relocation was discussed and the implications mentioned about stress, survival rate, and disease susceptibility. Relocated fish must be handled with care to prevent fungi and other external parasites from growing and parasitizing on host fish.

17 March   **Oribi Gorge Nature Reserve, South Coast**

I went out to allocate the game guards and conservancy guards according to their rostered positions along the series of dams in use during the fishing competition. I again assisted with the weighing in of fish, separating dead and live fish, and relocating the live ones into an aerated tank before dam restocking could take place. A few, additional licenses were issued as well as checking on the anglers. The event was concluded by a prize-giving ceremony, which I attended.

18 March   **Oribi Gorge Nature Reserve, South Coast**

We went out to shoot a sick farm horse in the conservancy upon request of its owner, and scattered the carcass at a vulture restaurant (feeding spot) on top of a hill. Returning to home base, the labourers were dropped at the picnic spots to do some grass cutting and general outdoor maintenance. We then went to the Port Shepstone/Marburg area to place an advert in the South Coast Herald so as to increase tourist influx to Oribi Gorge Reserve. A stepladder and set of lawnmower blades were also obtained from various outlets. Afterwards, the labourers were picked up again.

19 March   **Oribi Gorge Nature Reserve, South Coast**

We went off-base to observe all the game guards on a training day held for the Umtamvuna district. The programme consisted of reinforced military basics (marching), assessment on previous educational video regarding conservation, and general patrolling of the area to search out snares. The day was ended by various soccer matches that were played between teams. A conservation talk was given by the zone officer (district warden), before twenty-one year service bars were presented to a senior game guard.

20 March   **Oribi Gorge Nature Reserve, South Coast**

I had gone through the zone officer's manual and law book in order to gain a better understanding of law enforcement practices pertaining to the Nature Conservation Ordinance. Later in the day, I went out with the regular Banana Express tourist group to do the Baboon View interpretive trail and basic plant identification as per individual request.

21 March   **Oribi Gorge Nature Reserve, South Coast**

We went to Port Shepstone's lower regional court because of a case of bark-stripping – a violation of the Nature Conservation Ordinance. The accused might also have been charged with trespassing. Allegedly, he was caught on the fence boundary which complicated matters. Because the accused wasn't detained or out on bail, he didn't turn up and a warrant for his arrest was issued. On the way back, a number of cold drinks were bought at a Coca Cola factory for the rest camp's shop. Labourers were picked up from the zone officer's official residence, where they had cut the grass and tended the garden. Game guards were left behind to run back to base as a training exercise.

22 March   Oribi Gorge Nature Reserve, South Coast

We went to the vulture restaurant to check if the horse carcass of a few days ago had been preyed upon. To our surprise, six vultures were identified feeding. We subsequently set out for a nearby farm to patrol an area suspected of poaching activities. Although seven snares were uncovered near the fence boundary, no superficial dwellings or fireplaces were come across as might have been expected.

23 March   Oribi Gorge Nature Reserve, South Coast

We headed out to another farm nearby for the collecting of a dead cow and spreading the carcass at the Hell's Gate vulture restaurant. I later went out with the Banana Express visitors, as usual, on the Baboon View interpretive trail. Conservation practices throughout Africa, but especially those of Zambia and Zimbabwe, were discussed.

24 March   Oribi Gorge Nature Reserve, South Coast

I went out for general patrols in order to uncover any snares, feral dogs, poachers or illegal activities, but no such violations were found. A vast area was patrolled, including Hoopoe Falls, View Trail and Nkonka Point.

25 March   Oribi Gorge Nature Reserve, South Coast

Because of excessive rain, construction on the Hoopoe Falls trail was suspended and workshop jobs were allocated instead. The sliding doors of the gents' shower in the rest camp were fixed, as they were dislodged due to overuse by tourists. Basic plant identification was done in discussion with a horticulturist employed by the Board who had been on vacation at Oribi Gorge.

26 March   Oribi Gorge Nature Reserve, South Coast

I went out with the labourers to the Hoopoe Falls trail construction site so as to continue with the making of the path. Recommendations were, firstly, to prevent fragmentation of big rocks in order to preserve the geological appeal of the visible landscape (outcrops), and secondly, to minimize destruction of the surrounding plant life as far as possible.

27 March   Oribi Gorge Nature Reserve, South Coast

Official Day Off (spent at Ballito, Dolphin/North Coast).

28 March   Oribi Gorge Nature Reserve, South Coast

Official Day Off (spent at Ballito, Dolphin/North Coast).

29 March   Oribi Gorge Nature Reserve, South Coast

Official Day Off (spent at Ballito, Dolphin/North Coast).

30 March   Oribi Gorge Nature Reserve, South Coast

Official Day Off (spent at Ballito, Dolphin/North Coast).

31 March   Oribi Gorge Nature Reserve, South Coast

Official Day Off (spent at Ballito, Dolphin/North Coast).

Date: 23/03/91   Station: ORIBI GORGE

Activities/Observations:

WENT OUT TO MR BARRET'S FARM FOR THE COLLECTING OF A DEAD COW AND THE PLACING AT VULTURE RESTAURANT (OPENING OF CARCASS ETC.)

WENT OUT WITH BANANA EXPRESS VISITORS FOR BABOON VIEW INTERPRETIVE TRAIL. CONSERVATION TROUGHOUT AFRICA, ESSPECIALLY ZAMBIA AND ZIMBABWE WAS DISCUSSED.

APRIL

1 April Rhino Bomas, Umfolozi Game Reserve

Upon arrival at the new station, I acquainted myself with immobilizing agents, darts, needles, accessories as well as policy and procedures concerning game capture. Basic administration for every day was considered as well as the strategy to allocate captured animals. Differences between white (square-lipped) rhino and black (hook-lipped) rhino were noted in terms of capture strategies and containment in bomas (heavy-duty holding pens). Infrastructure in progress was observed and future building and general maintenance plans were discussed with the officer-in-charge. Ear-tagging of rhinos for auctioning purposes was noticed. The feeding of captured animals (daily rations) and cleaning routine of the bomas were also under discussion.

2 April Rhino Bomas, Umfolozi Game Reserve

Arrangements for departure by helicopter and Land Rovers were discussed and the details thereof deliberated. Immobilizing darts, for example, were made up. Rhino offloading was at the order of the day for different allocations. Later on, roofing was applied to one of the boma shelters under construction. Note was taken of other onsite jobs, such as fence construction (hole-digging and alignment of posts), feeding regime and overall modus operandi.

3 April Rhino Bomas, Umfolozi Game Reserve

I went out with the rhino capture team. The operation's vehicle fleet consisted of a helicopter, heavy-duty trucks, Land Rovers, and a fuel car. A specialized approach to procedures was noticeable. Searching and darting was done by the helicopter crew, while the ground crew pursued by vehicle through the bush. The veterinarian followed up with the administering of antibiotics and nalorphine compound, after which the capture team could guide the recovering but still drowsy rhinoceros into its reinforced crate for uploading by winch onto the back of the truck. Subsequently, departure for offloading at the rhino bomas took place. Later in the day, I continued to give roofing assistance to the building foreman in order to complete another rhino shelter. We started breaking down a rotten roof of an existing shelter due for replacement.

4 April Rhino Bomas, Umfolozi Game Reserve

We continued with the construction of roofs. Holes were dug for extra poles. Base beams of adjacent roofs were erected in a stepping style consistent with the slope of the ground. I occasionally checked on the local construction of giraffe bomas, where use was made of taller poles than for rhino bomas. I also checked on hole-digging along the new fence line. Rectangular holes were dug at right angles to the fence for ease of lining up the posts.

5 April   **Rhino Bomas, Umfolozi Game Reserve**

I assisted with the translocation of rhino, moving them from boma to gate-docked cage (crate) to truck for transport to their new destination. The different techniques and methods used to achieve that were considerable. Process in short: darting; caging; winching onto heavy-duty truck; transferring to 25-ton truck capable of transporting six rhinos in separate compartments at the same time. Fodder was provided, but no water was necessary because of the short journey. Once the truck departed, we continued with construction to the giraffe cages and fences. Temporary living quarters were removed because of practical and aesthetic reasons, and then re-erected near the existing staff compound.

6 April   **Rhino Bomas, Umfolozi Game Reserve**

Several tasks at different construction sites were completed: Giraffe bomas' fence (logs) on one side; concreting in of extra poles to support roof of shelter; concreting in of post poles on fence line. Different capture techniques, methods and strategies concerning even-toed ungulates were discussed. We ended the day by filling the gas cylinders for domestic purposes.

7 April   **Rhino Bomas, Umfolozi Game Reserve**

Construction work continued for the different tasks at hand; these included giraffe cages, shelter roofs, and fences. A general cleanup was done afterwards. We went out into the field to cut grass for hay/fodder. A routine check-up was done later on to ensure the rhinos (especially newly-captured) took their food and defecated. General appearance was also noted in case of illness. Wilderness trail objectives and activities were discussed as well as the lasting effects of the Domoina flood on Mdindini Camp and surrounding landscape.

8 April   **Rhino Bomas, Umfolozi Game Reserve**

Pole erecting for the inner fence was continued. Giraffe cage construction was continued by means of trench-digging and pole installation. The old fence was dismantled – cables and poles were recycled. Meanwhile, general tasks like routine feeding and cage cleaning continued almost unawares. Feeding strategy involved green, freshly-cut grass supplemented by vitamin B-complex and equine cubes for newly-captured rhinos. Once adjusted (shortly held), they were given hay and lucerne also but no supplements. The well-adjusted (longer held) rhinos were fed on hay only. Holding setup per individual rhino consisted of a large boma for roaming, linked to a smaller, inner boma, exiting into a cage docked for transporting purposes.* Immobilizing agents used were M99 and hyoscine added together in sterile water, with nalorphine and M5050 antagonists as the reversal.

* See field diagram on p.27 for how feeding and caging dovetail each other rotationally (i-iii)

9 April   **Rhino Bomas, Umfolozi Game Reserve**

A routine check around the bomas was done on the rhinos' welfare and their fodder given. Giraffe boma construction continued in the background. Horizontal support poles were attached on the one side. Fence operations included bush clearance (minimally done) along the fence line so as to create and maintain a patrol path. Fascia boards and purloins were fixed to the support poles of a roof shelter. I ended the day by reading Appendix 6 of 'Management of Wilderness Areas.'

10 April   **Rhino Bomas, Umfolozi Game Reserve**

No irregularities were found during the routine check-up, except for a burst water pipe connection which was fixed. Construction work continued on fences, bomas, and shelter roofs. Vertical poles were attached to the horizontal bars of the boma under construction. Asbestos Everite sheets were fitted to the roof frame (beams and purloins), using roofing screws, and so another shelter was completed.

11 April   **Rhino Bomas, Umfolozi Game Reserve**

I took an official vehicle to the service station for general servicing, especially wheel balancing and alignment. I subsequently went to Hilltop camp at Hluhluwe Game Reserve and observed the veterinary treatment of a captured springbok. It was tranquilized by a phrenilin-filled dart, projected through a blowpipe. After the injured antelope's horn was amputated, it was given antibiotics. Back at base, an outer wall of a boma was reinforced because a female rhinoceros and her calf had pushed it over and escaped the previous night. Elsewhere on the premises, a broken roof shelter, due to a rhino using the support poles as rubbing posts, was fixed and strengthened by means of attachment bars to the boma's sidewall. Holes for bolts were drilled for reinforcing the timber structures of shelters.

12 April   **Rhino Bomas, Umfolozi Game Reserve**

A routine check-up was performed with the officer-in-charge and the veterinarian – no problems were identified. Construction at different sites was continued. Further poles were erected as posts according to ground level to form another side of the giraffe boma. Old shelter timber was broken down and measurements for new timber were taken. Bolts for reinforcement were made by grinding away the edges of iron rods to taper, so that thread could be cut by a stock-and-die set.

13 April   **Rhino Bomas, Umfolozi Game Reserve**

We conducted a routine check around the rhino bomas: no apparent problems. The usual work at construction sites was done. Beams and purloins as well as other supports were installed. Bolts to other shelter roofs were also fitted. Different plans of post placement for fence construction were discussed so as to find the best design for strong, long-lasting anchorage that is still cost-effective. (Exposed anchor cables may cause injury to animals) *p.28*

14 April   **Rhino Bomas, Umfolozi Game Reserve**

We did a routine check-up and the condition of the rhinos was found to be in order. Different construction work continued like the nailing of asbestos sheets to the roofing timber, thereby completing yet another boma shelter.

15 April   **Rhino Bomas, Umfolozi Game Reserve**

A routine check on the rhinos was done, as for every morning, and no irregularities were noticed. Further bolts were made, a hammer handle was fixed, and a pipe was welded. I also fetched an official vehicle from Hluhluwe Service Station, as it had been serviced and repaired. Fence construction methods for ideal anchorage of posts were discussed again with the officer-in-charge, bearing in mind all the implications and practicalities.

16 April   **Rhino Bomas, Umfolozi Game Reserve**

The routine check-up was done. It was decided that two juvenile rhinos ought to get more lucerne than the status quo recommends, fulfilling in their higher protein demand (thus 25% lucerne and 75% teff hay). We subsequently went to Mtubatuba Railway Station to collect a lucerne load which was, back at base, packed into the shed. Measurements for another boma shelter and excavation of holes for its poles were started. Procedures to comply with building requirements were discussed with regard to foundations. Usual construction work continued around the premises.

17 April   **Rhino Bomas, Umfolozi Game Reserve**

I did a routine check-up on the rhinos and everything was found to be alright. Further construction work was carried out. I observed the putting up of a fence by the use of a pulley to erect the wire under tension. Setting of poles was done after excavation and concreting according to measurement, levelling and alignment.

18 April   **Rhino Bomas, Umfolozi Game Reserve**

I did the routine check-up and everything was fine. I then went to fetch the Land Rover that had been in for general servicing at Hluhluwe Service Station. Back in Umfolozi, the usual construction was done at different sites. Gates were set for fitment in order to allow controlled access to the staff compounds from the pens/bomas. Beams were fitted as well as fascia boards pertaining to a shelter roof.

19 April   Rhino Bomas, Umfolozi Game Reserve

A routine check-up on the rhinos revealed nothing out of the ordinary. Usual construction projects continued around the premises. Work on the shelter roof involved purloins being fitted, while work on the fence involved gates being fitted. The National Diploma in Nature Conservation programme was discussed with the officer-in-charge, sharing ideas on its implications for the field and its level of adequacy pragmatically. Lastly, a dilapidated reed roof alongside the office/squaredavel (my living quarters) was demolished.

20 April   Rhino Bomas, Umfolozi Game Reserve

I checked up on the rhinos around the bomas as was done every day. The usual construction work was also continued at various sites. Later on, corrugated asbestos sheets were nailed down to complete another roof shelter.

21 April   Rhino Bomas, Umfolozi Game Reserve

I did a routine check-up first thing in the morning and everything was fine. The usual construction took place around the bomas. The finishing touches were applied to a recently completed roof shelter by cutting off overhung pieces of beam. The making of new bolts was also done. We started digging holes for the poles of a new rhino shelter to come. I later filled up the petrol tanks of the Land Rovers as well as a container with fuel for the generator. I replaced a spare wheel with a newer treaded one.

22 April   Rhino Bomas, Umfolozi Game Reserve

A routine check-up on the rhinos was done while the usual work at construction sites commenced. Later in the morning, I went to a physician for an anti-tetanus injection. We then departed for Ithala Game Reserve to assist with their elephant reintroduction programme.

23 April   Rhino Bomas, Umfolozi Game Reserve

After I had arrived back from Ithala, I continued assisting with construction work to fences, roofs, bomas and cages. Bluegum poles for the new rhino shelter were set in concrete.

24 April   Rhino Bomas, Umfolozi Game Reserve

I did a routine check-up and everything was found to be satisfactory with the rhinos. The usual construction was done at different sites around the bomas. Beams and fascia boards were fitted to the poles of the roof shelter.

25 April　　　　　　　　　　　　Rhino Bomas, Umfolozi Game Reserve

I started the daily routine by checking up on the rhinos. A combination of agriflavine and glycerine was sprayed on superficial wounds to prevent septicaemia and encourage skin rejuvenation. Bactidol was used to kill fleas and their eggs, preventing penetration and degradation of wound tissue. The usual construction work continued. I checked the fluids and key mechanical parts of the vehicles which would be used for game capture operations the next day. I ended the day by investigating the water provision system of the rhino bomas. This was composed of a series of supply channels interconnecting the various bomas.

26 April　　　　　　　　　　　　Rhino Bomas, Umfolozi Game Reserve

Immobilizing darts for the capture of adult white rhino were made up in particular concentrations of M99, fentanyl, hyoscine, and sterile water. I set up the ground reserves for helicopter refuelling at base and aboard the petrol truck to support the aerial operation. The necessary hand pump was also fitted. A rhinoceros – once darted from the air – was pursued through the bush by several all-terrain vehicles. After coming upon the sedated rhino, an ear tag was inserted for identification purposes. A local antibiotic was administered into the dart wound and general antibiotics were given intramuscularly. Local wounds were treated with Airbiotic GV (gentian violet) spray. In the case of female rhinos, a blood sample was taken to test for pregnancy. The antidote compound given intravenously in one ear was followed up by putting a grain bag (blindfold calming effect) and thick rope around the head and horns of the rhino, which aided us in guide-walking the animal into a reinforced crate. This was then winch-loaded onto a heavy-duty rollback truck for translocation.

27 April　　　　　　　　　　　　Rhino Bomas, Umfolozi Game Reserve

I checked the vehicles and made ready for another rhino capture operation. I also organized aircraft fuel and drove the fuel vehicle, waiting in accessible places for helicopter landing and refuelling. The different methods, techniques and strategies unique to rhino capture were observed. Analysis of blood tests was discussed. Amounts of progesterone in female rhinos indicated state and stage of pregnancy.

28 April　　　　　　　　　　　　Rhino Bomas, Umfolozi Game Reserve

I organized fuel for the helicopter and followed in the fuel vehicle. We specifically looked for unmarked (not part of research programme) and relatively young black rhino adults with a limited horn length. Immobilizing agents were different and comprised particular concentrations of M99, fentanyl, azaperone, and sterile water. It must be stressed that black rhino, being less densely distributed, were captured across the entire area of the reserve, but white rhino (even though more abundant) could only be taken from specified zones.

29 April Rhino Bomas, Umfolozi Game Reserve

I organized fuel for the helicopter and vehicles. Game capture eventuated as per usual procedure. Two white and one black rhino were captured. The tip of the front horn of the more aggressive black rhino was cut off because of injuries likely caused once released. Back at the bomas, a cyst underneath a rhino's main horn was punctured to release the puss. The collapsed cavity was injected with peroxide and iodine solutions, followed by an application of Stockholm tar. Penicillin was also administered, although chances were that the horn would be lost. Doxapram was administered to maintain respiratory function while under M99 sedation (antidote: M5050). I subsequently went to the Hluhluwe pens to collect new aircraft fuel in exchange for empty drums. On my way back, I picked up a parcel at Hluhluwe Service Station for the pilot.

30 April Rhino Bomas, Umfolozi Game Reserve

I organized the fuel (avgas) reserves and fuel vehicle as ground support to the game capture operation. One black rhino was captured. Afterwards, general neatening of the boma area was discussed in preparation of the annual game auction. Usual construction work was also done. I took an automotive technician to the scene where a capture truck had broken down. It was discovered that a pressure valve on the brake system was broken. After the technician had fixed the valve, we couldn't manage to get it out of the sandy soil. Another heavy-duty truck would be dispatched to tow it unstuck.

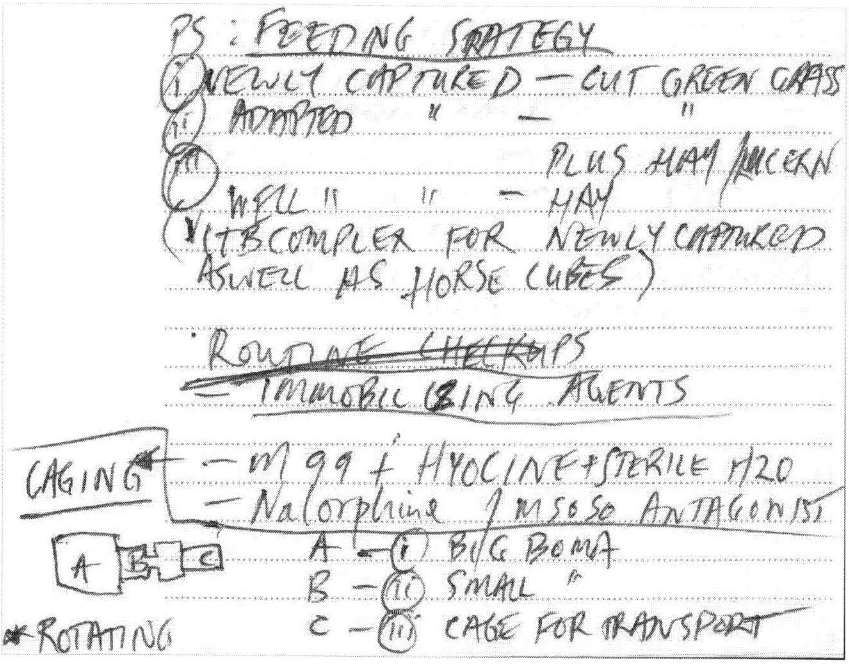

[Handwritten notes:]

PS: FEEDING STRATEGY
(i) NEWLY CAPTURED — CUT GREEN GRASS
(ii) ADAPTED   "   —   "   PLUS HAY LUCERN
(iii) FULL   "   — HAY
(TB COMPLEX FOR NEWLY CAPTURED AS WELL AS HORSE CUBES)

ROUTINE CHECKUPS
IMMOBILIZING AGENTS
— M99 + HYOCINE + STERILE H2O
— Nalorphine / M5050 ANTAGONIST

CAGING
A — (i) BIG BOMA
B — (ii) SMALL   "
C — (iii) CAGE FOR TRANSPORT

*ROTATING

Date: 30/04/91   Station: UMFOLOZI BOMAS
Activities/Observations:
ORGANISED FUEL (AVGAS) =

Date: 13/04/91   Station: UMFOLOZI BOMAS
Activities/Observations:

→ AS DONE

→ GOOD

→ BEST

A NEW Hunts individuals → BETTER

AND LOADING WITH WINCH
OR BY COCKING IT BACK-
WARDS AND FORWARDS.
BY REVERSING (ROCKING).

Steven Ward/Rhino Capt

MAY

1 May   Rhino Bomas, Umfolozi Game Reserve

I went along to the scene of the previous day's breakdown in a heavy-duty truck which was all-wheel drive. We managed to pull the repaired but stuck truck free. Meanwhile, the usual construction work continued back at the bomas. Preparations were made for the casting of a concrete floor underneath a shelter roof. I later brought a new caravan from another site to base camp, and took the existing one to the other site (both were levelled for occupancy). I also erected a side tent onto the new caravan and laid out a groundsheet. Vehicles from the capture operation were being washed and made ready for standby.

2 May   Rhino Bomas, Umfolozi Game Reserve

Three juvenile rhinos were loaded for an exhibition. Usual construction work continued. Final adjustments to the caravan tent were made. The grass surrounding the caravan was cut. Later in the day, we started to move old crate frames from their existing positions to an area immediately out of sight. These frames were steel-brushed to remove rust and then painted with a metal primer. Any remaining planks were removed and old bolts retrieved where possible. I had some gas bottles filled, fetched clean linen, and dropped the monthly administration work of the clerk off at the typist/secretary. I also fetched petrol for the portable welding plant. Upon my return, the remainder of the game capture vehicle fleet was checked and washed.

3 May   Rhino Bomas, Umfolozi Game Reserve

We took the new Land Rover to Hluhluwe Service Station, as the conservator was not satisfied with the job done to the front axle/steering gear (clutch was defected too). On the way back, I dropped the driver at Mtubatuba to do some shopping. After having arrived at the bomas, I went further with crate assembling. Planks were cut to the correct size and holes drilled for bolting onto the steel frame. Other frames were primed and prepared for new planks. The usual construction continued elsewhere on the premises.

4 May   Rhino Bomas, Umfolozi Game Reserve

I continued with crate assembly. Old bolts from rotten planks were reused for attaching the new planks. Planks were cut by using an electric saw. We experienced difficulty with the power generators, but no obvious fuel or electrical problem could be found. One generator wouldn't start and though the other one started, it didn't generate sufficient power.

5 May Rhino Bomas, Umfolozi Game Reserve

I continued to work on the rhino crates, stripping the old planks off the steel frames for reuse and subsequently cutting new planks to the required lengths using the handheld circular saw. I also assisted in searching for technical faults on the power generators. It appeared that a sparkplug was defective on the one and the brushes were worn on the other.

6 May Rhino Bomas, Umfolozi Game Reserve

Official Day Off (spent at base).

7 May Rhino Bomas, Umfolozi Game Reserve

Official Day Off (spent at base).

8 May Rhino Bomas, Umfolozi Game Reserve

Official Day Off (spent at base).

9 May Rhino Bomas, Umfolozi Game Reserve

Official Day Off (spent at base).

10 May Rhino Bomas, Umfolozi Game Reserve

Official Day Off (spent at base).

11 May Rhino Bomas, Umfolozi Game Reserve

I continued with the assembly of rhino crates. Fresh planks were cut according to specific requirements for each crate by means of electric saw.

12 May Rhino Bomas, Umfolozi Game Reserve

I did more construction on rhino crates. Assembly entailed planks being cut to specified size and bolted to primed, steel frames.

13 May   Rhino Bomas, Umfolozi Game Reserve

Rhino crate production continued. Old crates were dismantled by means of a cutting torch controlled by a mixture of acetylene and oxygen. More planks arrived and were offloaded for fitment to the prepared, steel frames.

14 May   Rhino Bomas, Umfolozi Game Reserve

We continued dismantling more of the old rhino crates. The stripped frames were painted with primer after rust removal. In between, I ripped some planks for a roofing job at the bomas. I resumed cutting planks for the crates to frame size and started assembling them. I ended the day's work by painting the giraffe bomas' gate not yet hung at the time.

15 May   Rhino Bomas, Umfolozi Game Reserve

I made further progress on rhino crate assembly: bolting of planks after having been cut or ripped to size; painting of more steel frames with primer; etc. We went out into the bush to cut fresh, green grass for the newly-arrived white rhinos which had not yet adapted to artificial feed, such as teff, lucerne and supplementary cubes.

16 May   Rhino Bomas, Umfolozi Game Reserve

I carried on assembling additional rhino crates. Planks for fascia boards of shelter roofs were also ripped to size. We painted some more crate frames with primer for steel surfaces in the afternoon.

17 May   Rhino Bomas, Umfolozi Game Reserve

I went further with crate construction: plank cutting/ripping to size and bolting to painted frames. Seven crates had been done up to date, with only the doors remaining to be tailored and fitted to each. Later in the day, I went to fetch fuel for the welding plant.

18 May   Rhino Bomas, Umfolozi Game Reserve

We continued with rhino crate assembly as per usual.

19 May   Rhino Bomas, Umfolozi Game Reserve

I continued further with the construction of rhino crates. Nine crates (without doors) were completed thus far. We carried on applying primer to bare frames and started dismantling more old crates for frame preparation and cladding with new planks.

20 May   Rhino Bomas, Umfolozi Game Reserve

We went further with rhino crate construction. I used a cutting torch in order to finish detaching the old, bolted planks from their frames which were then primed. Outstanding spots on the new crates were also touched up with the paintbrush. I ended the day's duties by ripping more planks for beams/fascia boards of shelter roofs.

21 May   Rhino Bomas, Umfolozi Game Reserve

I continued with crate construction. We started to attach the wood to the doorframes after having been painted with primer. More planks were ripped by circular saw for roofing purposes (rhino shelters).

22 May   Rhino Bomas, Umfolozi Game Reserve

I fitted the doors of a rhino crate (upper section) and went further with dismantling old crates and painting frames. We went to Empangeni/Richards Bay to fetch a repaired vehicle from the panelbeaters (Auto Cars) and left the bigger truck at the Mercedes dealer for propshaft repairs. I also paid for the previous repairs done to the truck's pressure valve. On our way back, I collected a towing winch cable at Haggie Rand.

23 May   Rhino Bomas, Umfolozi Game Reserve

I continued with rhino crate construction. Large holes were drilled in the crate's floor for steel pipes to be inserted, forming vertical bars in the front and back. The two, middle pipes in the front had been twisted to taper outwards at the bottom in order to allow feeding.* The remaining crate doors, which had been completed the day before, were hung. Prior to assembly, paint was applied to frames and pipes alike. I also ripped some planks for fascia-beams of shelter roofs.   *See field diagram on p.34 for clarity*

24 May   Rhino Bomas, Umfolozi Game Reserve

I went further with crate construction by assembling pipes after holes had been drilled through floors and by attaching wood to doorframes. I finished the day by ripping more planks.

25 May   Rhino Bomas, Umfolozi Game Reserve

I continued with the construction of rhino crates. Doorframes were modified by cutting torch for better fitment, after I had welded on some hinges. We repainted any scorched areas as a result of the welding and torching. Doors were assembled and fitted to the crate.

26 May   Rhino Bomas, Umfolozi Game Reserve

Crate construction entailed the painting of pipes (vertical front and rear bars) before insertion, making of doors (after dismantling old ones), and ripping of new planks. I subsequently went offsite to witness a giraffe capture operation, noticing the temporary construct of a funnel (sheet walls strung up by cables) as well as specialized aspects of catching strategy, methods and techniques used.

27 May   Rhino Bomas, Umfolozi Game Reserve

Crate construction, in particular door assembly and fitment, continued. I drilled 51mm-diameter holes in crate floors for pipes (vertical bars), reaming some wider for bigger pipes. Holes were also made through doorframes by cutting torch for bolt insertion.

28 May   Rhino Bomas, Umfolozi Game Reserve

Crate construction continued – hole-drilling, pipe painting and assembly. We also dismantled more of the old crates. I had cut some planks for crate doors to the required size and started assembling them, before going to Empangeni to fetch a game capture truck from Mercron. Unfortunately, it hadn't been fixed, as the propshaft part was too long. We had to turn back empty handed. Apparently, a message was left at Game Capture Headquarters (Hluhluwe) concerning this.

29 May   Rhino Bomas, Umfolozi Game Reserve

Crate construction continued as I made holes for pipes with the holesaw drill. More old crates were dismantled. I welded on some broken-off hinges to the doorframes of a crate as well as a winch ring onto its chassis. Doorframes were also modified where necessary.

30 May   Rhino Bomas, Umfolozi Game Reserve

Crate construction – door assembling/fitting, painting and pipe fitment continued. Abandoned rhino trailers used under a previous capture regime were towed away, bulldozed into a hole in the ground and filled over with soil.

31 May   Rhino Bomas, Umfolozi Game Reserve

We carried on with rhino crate construction in terms of pipe fitment, door assembly and painting. I also made the necessary holes for bolts through steel frames by means of a cutting torch.

Date: 23/05/91   Station: UMF02021 Bombas

Activities/Observations:

WENT FURTHER WITH CRATE CONSTRUCTION: DRILLED HOLES FOR PIPES AND ASSEMBLED PIPES. DOORS WERE HANG AS FINISHED YESTERDAY. NECESSARY PAINT GIVEN. RIPPED PLANKS FOR FASHIA/BEAMS.

FRONT
UPPER DOOR
LOWER "
BACK

JUNE

1 June   Rhino Bomas, Umfolozi Game Reserve

Rhino crate construction continued. We assembled more crate doors by drilling and bolting of planks, after I had cut them according to the specified requirements.

2 June   Rhino Bomas, Umfolozi Game Reserve

The construction of rhino crates progressed as we persisted in assembling doors. Planks were cut by circular saw to the required size and all steel surfaces were painted. I also welded some working utensils which had been broken.

3 June   Rhino Bomas, Umfolozi Game Reserve

Crate construction continued in terms of door assembly and fitment, with the necessary cutting-torch modifications, bolting and painting applied. Crates were numbered according to size: #4 (any juvenile rhino); #3 (adult black rhino); #2 (smaller white rhino); #1 (larger white rhino).

4 June   Rhino Bomas, Umfolozi Game Reserve

Construction of rhino crates in terms of door assembling, fitting, welding, and painting continued. We also assembled the floor planks of a new crate.

5 June   Rhino Bomas, Umfolozi Game Reserve

We went further with rhino crate construction by assembling planks to the painted frames, after I had cut and ripped the planks to size. I also welded a broken-off towing ring onto a crate's frame and cut thick, fence cables with the cutting torch.

6 June   Rhino Bomas, Umfolozi Game Reserve

Crate construction continued as we dismantled more, old crate doors and prepared them for fitment. Plank measuring/cutting, frame painting, large-diameter drilling and pipe fitting were at the order of the day. I also fixed a broken hammer.

7 June   Rhino Bomas, Umfolozi Game Reserve

Crate construction continued in terms of plank cutting/ripping and door assembling/fitting. I had also done some welding. Later on, I assisted with the offloading of a sub-adult white rhino at the bomas. The working of the electric motor was discussed, in particular the armature, field coil and brushes, while the brushes of a drill were being replaced by the officer-in-charge.

8 June   Rhino Bomas, Umfolozi Game Reserve

Construction of rhino crates continued as the completed doors were fitted and more old crates dismantled for restoration. The stripped frames were painted with a metal primer.

9 June   Rhino Bomas, Umfolozi Game Reserve

I pressed on with rhino crate construction by plank cutting and ripping for floor assembly. Repairs to an older crate were also performed (pipe painting and insertion). I did some welding on a 4lb-hammer to fix it.

10 June   Rhino Bomas, Umfolozi Game Reserve

Rhino crate construction involved frame preparation (painting) for cladding. We then assembled the ripped planks to the frame and doors.

11 June   Rhino Bomas, Umfolozi Game Reserve

We continued with door assembly using cut and ripped planks (incl. fitting of doors) as well as floor drilling for insertion of painted pipes, thereby finishing more rhino crates.

12 June   Rhino Bomas, Umfolozi Game Reserve

Crate construction continued in terms of assembling the doors and fitting them. Drilling of the crate floor for the fitment of painted pipes was carried out. I also repaired a damaged crate by welding in extra floor supports.

13 June   Rhino Bomas, Umfolozi Game Reserve

We finished all rhino crate construction for the game auction. Repairs to some crates were also performed. Auction arrangements were discussed in outline as well as the magnitude of hosting such an event each year.

14 June   **Rhino Bomas, Umfolozi Game Reserve**

General cleaning up of bomas and surrounding area (vast site) was done in final preparation of the upcoming game auction. I took an officer to the Hluhluwe pump station where he met up with a transport fleet of National Parks Board in order to accompany them to Skukuza.

15 June   **Rhino Bomas, Umfolozi Game Reserve**

Aspirant rhino buyers turned up at the bomas to view the lots and make their bidding choice for Auction Day – June 17th. I took the labourers to Nkangala in order for them to spend their official days off at home. Back at base, leaking water pipes were fixed by joining in of new sections. Male thread had to be cut to accommodate new fittings.

16 June   **Rhino Bomas, Umfolozi Game Reserve**

I assisted with the pre-auction activities. Capture sequence and techniques for white rhino and black rhino were also discussed.

17 June   **Rhino Bomas, Umfolozi Game Reserve**

I attended the long-awaited game auction. The price range of the different species (lots) varied notably. Overall, the auction had gone well and the venue was excellent. A black rhino didn't sell as high above the reserve as was expected, based on previous years. Legislative and veterinary restrictions on translocation were considered a contributing factor.

18 June   **Rhino Bomas, Umfolozi Game Reserve**

I went to Hilltop at Hluhluwe Game Reserve to drop off the temporary signboards which had been directing auction goers for the last few days. In addition, I was supposed to fetch a jackal from Underberg and drop two truck drivers in Pietermaritzburg on my way, but the arrangement was cancelled due to unforeseen circumstances. After returning to the bomas, I brazed together the loose wires on a leopard cage. I also welded a chevron onto the rear end of a capture truck.

19 June   **Rhino Bomas, Umfolozi Game Reserve**

Crate construction was resumed and more plank ripping and cutting to requirements ensued. We continued to paint crate frames before lining them out with the planks. Dismantling of old crates took place too. I also fetched petrol for the welding plant.

20 June	Rhino Bomas, Umfolozi Game Reserve

The additional rhino crates were completed by fitting pipes and doors and ensuring their proper painting. I also did some brazing and welding. Lastly, I assisted with the loading of a rhinoceros for relocation on delivery.

21 June	Rhino Bomas, Umfolozi Game Reserve

Official Day Off (spent at Ballito, Dolphin/North Coast).

22 June	Rhino Bomas, Umfolozi Game Reserve

Official Day Off (spent at Ballito, Dolphin/North Coast).

23 June	Rhino Bomas, Umfolozi Game Reserve

Official Day Off (spent at Ballito, Dolphin/North Coast).

24 June	Rhino Bomas, Umfolozi Game Reserve

Official Day Off (spent at Ballito, Dolphin/North Coast).

25 June	Rhino Bomas, Umfolozi Game Reserve

Official Day Off (spent at Ballito, Dolphin/North Coast).

26 June	Rhino Bomas, Umfolozi Game Reserve

Official Day Off (spent at Ballito, Dolphin/North Coast).

27 June	Rhino Bomas, Umfolozi Game Reserve

Official Day Off (spent at Ballito, Dolphin/North Coast).

28 June	Rhino Bomas, Umfolozi Game Reserve

Official Day Off (spent at Ballito, Dolphin/North Coast).

29 June   Rhino Bomas, Umfolozi Game Reserve

Official Day Off (spent at Ballito, Dolphin/North Coast).

30 June   Rhino Bomas, Umfolozi Game Reserve

Official Day Off (spent at Ballito, Dolphin/North Coast).

Rhino capture underway in the Umfolozi.

White rhino calf with the author.

Block burning towards Injasuti camp in the Drakensberg.

Fire management in the Berg.

JULY

1 July   Monks Cowl Nature Reserve, Drakensberg

Official Day Off (general acquaintance).

2 July   Monks Cowl Nature Reserve, Drakensberg

Official Day Off (general acquaintance).

3 July   Monks Cowl Nature Reserve, Drakensberg

Official Day Off (general acquaintance).

4 July   Monks Cowl Nature Reserve, Drakensberg

I continued to get acquainted with my new station by studying the names, shapes, positions and directions of the surrounding peaks. We went out on patrol towards Injasuti camp to investigate the spread of an incidental wildfire. Firebreaks were made in order to block off any spread. The fire had crept up Cathkin Peak where it was likely to die out. The area burnt stretched all the way from the Sterkhorn vicinity down to the Delmhlwazini stream.

5 July   Monks Cowl Nature Reserve, Drakensberg

We did incident fire control by means of block burning from firebreaks. Diesel-drenched corncobs extended on handheld wires (for drawing a burn line), fire beaters, and water sprayers were the items used in the operation. Later on, footpath construction was discussed in terms of planning, water flow, hiker prevalence, anti-erosion measures, etc. Logs and stones were often used. We also patrolled the area from Sterkhorn to Keith's Bush, where a basic campsite was checked upon. Gray's Pass was explored as well.

6 July   Monks Cowl Nature Reserve, Drakensberg

We patrolled the area from Monks Cowl base to Culfargie Outpost via Verkykerskop. A snare was found and confiscated. Stable Cave was checked for occupants (permits required), but nobody was encountered. I frequently observed the mountain profile and took particular note of the different peaks. At Culfargie, the patrol horses were given a deworming treatment.

7 July   **Monks Cowl Nature Reserve, Drakensberg**

Deworming of the game guards' horses continued. I subsequently examined the minutes of previous annual meetings in order to form a rough idea of the local management strategy, objectives and achievements. Footpath construction and maintenance seemed to have been priority as well as erosion control. Firebreaks and controlled burns in area blocks were also high priority. We drafted a new strategic plan for burning, based on previous years.

8 July   **Monks Cowl Nature Reserve, Drakensberg**

We burned two blocks between Blind Man's Corner and Van Heyningen's Pass. Unfortunately, the fire jumped the Delmhlwazini stream which was the physical barrier. The fire subsequently jumped Ship's Prow Stream and was put out by back-burns from the east and west at Old Woman Stream nearby Old Woman Grinding Corn, not too far from Injasuti Hutted Camp (overtime: 1700-0400 hours).

9 July   **Monks Cowl Nature Reserve, Drakensberg**

I checked on the building improvements of living quarters in the staff compound, such as painting, door hanging, etc. Location for horse stables was also explored. I resumed my examination, analysis and interpretation of the block burn strategy. The mosaic pattern of burning was necessary to ensure habitat access for the displaced species. Roughly half the area of the reserve must be burnt due to low grazing pressure. Lastly, I familiarized myself with the two-way radio system of the station.

10 July   **Monks Cowl Nature Reserve, Drakensberg**

We put in a firebreak by means of trace-making, using Gramoxone herbicide along a line for follow-up burning, on the opposite side of the valley, taking Jacob's Ladder as the point of orientation.

11 July   **Monks Cowl Nature Reserve, Drakensberg**

We made firebreaks along the block boundaries of the Culfargie area. The effect of arson fires was used in our favour. Burning strategy encompassed trace-making (April/May), followed by firebreak (June/July) and block burning (August/September).

12 July   **Monks Cowl Nature Reserve, Drakensberg**

We furthered construction to the path at the foot of the mountain by fixing logs into the slope in such a way as to reduce water acceleration without damming up. On steeper slopes, rocks were used to add a stepping feature.

13 July   **Monks Cowl Nature Reserve, Drakensberg**

We checked the busy points on path routes going up the mountain from Drakensberg Sun via Steilberg and Grotto. Forty-two people were counted at Steilberg and four at Grotto. No illegal situations, such as horse or dog entry, permit violations or neglect of overnight registration, occurred. Cave permits were being issued only if people would like to make use of caves rather than tents. I also collected plant specimens for preservation in a press and identified some grasses and trees as a result.

14 July   **Monks Cowl Nature Reserve, Drakensberg**

We went out on horseback to patrol the fence line and adjacent area, stretching from base to near Injasuti camp, thus including Wonder Valley. No cattle crossed the reserve boundary as was often a problem. Two eland (large antelope) were spotted in the southern part of the valley, with Matterhorn to the northeast.

15 July   **Monks Cowl Nature Reserve, Drakensberg**

Labour was organized to finish work on the footpath. I went out on horse patrol with the game guards to police the area from base to Jacob's Ladder and KwaNdema. No violations occurred such as cattle, horses or dogs entering the reserve. Dog tracks and bare tree stems (due to bark stripping) were seen, but the perpetrators could not be found. Vast numbers of wattle trees were noticed along the watercourses (Makurumani, KwaNdema).

16 July   **Monks Cowl Nature Reserve, Drakensberg**

We resurfaced a steep vehicle track extending from the office to the workshop. Quarry was brought in and fragmented onto the track. The rocky surface was then smoothed over from one side to the other in a convex fashion to enable water runoff. Humps were also made to slow down the water flow and drain water from the road during rain.

17 July   **Monks Cowl Nature Reserve, Drakensberg**

I collected samples and identified trees in the native gardens of the campsite/office area for purposes of ordering National Tree List tags for visitor extension. I attended an onsite training session given by an official from Spioenkop Nature Reserve, explaining issues of general horse care and riding gear maintenance. Later in the day, a burst water pipe was fixed. I also assisted in issuing new overalls and safety boots to the labourers.

18 July　　　　　　　　　　　　　　　　Monks Cowl Nature Reserve, Drakensberg

Road construction of an off-ramp to accommodate a new vehicle inspection pit at the workshop commenced. First, we measured and started digging a trench for the inspection pit, large enough for a medium-sized truck. I later continued with plant identification.

19 July　　　　　　　　　　　　　　　　Monks Cowl Nature Reserve, Drakensberg

We resumed road construction at the workshop complex. Meanwhile, footpath construction continued in the gorge (Mpofana). I assisted with diesel requisites, etc. I also saw to renovation progress being made at the staff compound, such as treatment of moist walls by applying a combination of smoothed layers of acrylic primer, bonding paste, undercoat, and paint. Later in the day, three more eland were spotted in the northern Wonder Valley.

20 July　　　　　　　　　　　　　　　　Monks Cowl Nature Reserve, Drakensberg

I went to the northern bank of Sterkspruit to investigate new footpath options linkable with existing paths via the gorge. A lot of bush clearance would have had to be done, in order to access the sub-plateau. Multiple walks/loops up to the inner forests could have been considered, but such a route would have been too daunting for the average hiker to attempt. A cascade was discovered immediately below an existing path crossing over Sterkspruit; a possibility existed of that path extending lower down.

21 July　　　　　　　　　　　　　　　　Monks Cowl Nature Reserve, Drakensberg

We patrolled Wonder Valley on horseback from base, via Matterhorn, to Injasuti River. No irregularities were encountered. Game counted in the southern valley were common duiker (4); and in the northern valley common duiker (5), eland (5), mountain reedbuck (1), and grey rhebuck (2). We were called out to fight an accidental bushfire which broke out near Champagne Castle Hotel. Strategy was surface clearance along a line (emergency firebreak) as well as removing burnable material, such as pine and wattle debris, from the plantation floor (overtime: 0200-0500 hours the next morning).

22 July　　　　　　　　　　　　　　　　Monks Cowl Nature Reserve, Drakensberg

Road construction continued after a shallow-laden water pipe had been re-laid deeper in order to make possible the necessary drainage structures along the road. I oversaw the continuation of excavating a vehicle inspection pit. We also looked for a wandering dog in the campsite without success. If found it would have been chased out or destroyed and the owner sought, depending on certain circumstances.

23 July   Monks Cowl Nature Reserve, Drakensberg

Road and footpath construction continued. More logs were spaced out across the path and nailed down by iron hooks.

24 July   Monks Cowl Nature Reserve, Drakensberg

I dropped a pile of log poles near the path construction site by vehicle. I also dumped some soil quarried from the vehicle inspection pit at the road construction site. We went out to Wonder Valley to investigate an accidental fire, but upon closer inspection it was found to be under control (burning out against Spitsberg Stream). We later went out to the Champagne Castle property (adjacent to the reserve), as another arson fire had been lit in the pine forest. It, too, was found under control upon our arrival due to the efforts of the local hotel staff. Not long after that, though, we were called out again to yet another arson-lit fire at Champagne Castle Hotel, which on that occasion had reduced a barn to ashes. We assisted in patrolling the area in search of the suspect. As a result, a criminal case had been opened by the South African Police.

25 July   Monks Cowl Nature Reserve, Drakensberg

I took the game guards to do some shopping at Winterton, Bergville, and in Estcourt. Other official duties were carried out as well. A flat tyre was fixed and a new tube inserted.

26 July   Monks Cowl Nature Reserve, Drakensberg

I assisted in compiling a repair list for saddles, etc. I then went out to investigate the possibility of a footpath route on the northern side of Sterkspruit, on the second series of hills, higher above sea level than the previous investigations. We also went to the neighbouring Champagne Castle property to put out a forest fire. I subsequently fixed the lock on the backdoor of our official station wagon. We later went to Sikhaleni Senyoka near Culfargie in order to put out an arson/accidental fire. Lack of wind and presence of physical barriers in the landscape played a positive role in controlling the fire (overtime: 1700-0030 hours).

27 July   Monks Cowl Nature Reserve, Drakensberg

I went to Winterton and Bergville via Spioenkop Nature Reserve for official business as well as getting to the post office and bank on time. I was also buying confectionary for the camp's shop, PVC water pipe joining clamps, and fuel. After returning to base, I did the necessary office administration.

28 July   Monks Cowl Nature Reserve, Drakensberg

We patrolled Wonder Valley on foot from base to Matterhorn. Cattle dung was found, proving their illegal entry. However, no cattle were seen inside of the reserve boundary. Upon our return, I transported fence/building poles to Culfargie. A bale of hay for feeding the Monks Cowl horses was also picked up on the way back.

29 July   Giants Castle Nature Reserve, Drakensberg

I attended briefings on the overnight oribi game census, of which I was an official observer that year. A detailed map and recording chart of the operation were handed out to all involved.

30 July   Giants Castle Nature Reserve, Drakensberg

I did my part of the annual oribi count near Cold Ridge and recorded eland (3), blesbuck (6), and oribi (1). Afterwards, we walked up to Giants Hut to overnight before moving on to do the next day's coordinated count.

31 July   Giants Castle Nature Reserve, Drakensberg

I continued my oribi count in the high Berg at the foot of Giants Castle (peak) near the Loteni Jeep Track junction. I recorded blesbuck (incl. a juvenile) and eland. The count was called off due to bad weather. A debrief and analysis were presented by the scientist-in-charge, comparing data from previous years and trends of known animal groups. We subsequently investigated a biological sewage plant near the campsite, and were pleasantly surprised with its aesthetic design.

On horse patrol (Gatberg/Intunja in the background).

AUGUST

1 August   Giants Castle Nature Reserve, Drakensberg

I entered all data of the oribi counts for the past four years (1987-1991) into a computer and printed out hard copies for later analysis and interpretation in order to compile a concluding report. Data was derived from maps upon which sightings were plotted so that distribution patterns, migration routes and other inferences could be drawn there from. Due to bad weather, the second and last day of the counting programme was disregarded and rescheduled for a later date.

2 August   Monks Cowl Nature Reserve, Drakensberg

I went to Pietermaritzburg via Midmar Dam Nature Reserve in order to re-tyre two of our official vehicles. Gun safes were returned in town. General business was done, for instance, the return of one vehicle to the panelbeaters because of a dissatisfactory job.

3 August   Monks Cowl Nature Reserve, Drakensberg

Compound improvements, such as the filling of cracks in walls with paint filler/PVA mixture, resumed. I made basic electrical connections by fitting plugs to a cable for usage in the camp's shop. I then checked and prepared vehicles for emergency fire callout. Shortly thereafter, we went out to combat a fire raging near Champagne Castle Hotel on the northern side of the road. A private house was severely damaged in the fire. The area from Culfargie to Jacob's Ladder was blocked off in order for another fire to run via the lookout where the radio repeater was situated, which couldn't have been helped (overtime: 1700-0400 hours).

4 August   Monks Cowl Nature Reserve, Drakensberg

I fetched a vehicle from Culfargie for a ranger of Giants Castle Nature Reserve. I also paid tribute to the workforce at Culfargie with food rations for fire combating the previous night. We subsequently went to Makurumani spring to check on smouldering material in the vicinity, but it was found to be non-threatening (dying out). Fire was spotted at the top of Steilberg heading towards KwaHlathikulu forest, though that, too, seemed to be incapable of spreading. After returning to base, I filled up the vehicles' fuel tanks.

5 August   Monks Cowl Nature Reserve, Drakensberg

I organized labour for footpath construction. I also ensured that back-burning was done definitively to stop the threat of smouldering material flaring up at the foot of Steilberg, thus blocking off any conceivable spread. Back at the office, I mapped out the area burnt by recent wildfire (incl. arson). Later in the day, I walked a vast route of paths between the campsite, Sterkspruit and the plateau in order to draft a detailed map for tourists.

6 August   **Monks Cowl Nature Reserve, Drakensberg**

We did alien plant control in a hotspot area for wattle, where the trees were cut 15cm from ground level, the bark stripped and the exposed stems painted with 2% Garlon in diesel. After returning to base, I went to Pietermaritzburg for official business at Head Office. I also exchanged a door, picked up a mattress, a .303 rifle (on which repairs were not possible), firewood, general office equipment, and other appliances. On our way back, an official vehicle was picked up after completion of an improved panelbeating job.

7 August   **Monks Cowl Nature Reserve, Drakensberg**

Footpath construction and alien plant control continued. So did compound renovations. Painting of the interiorly-exposed asbestos roofs was done using road paint as a seal to prevent any contamination to occupants. I later checked and filled one of the reserve's vehicles with fuel. I also added to the tree tag list by identifying more plants.

8 August   **Monks Cowl Nature Reserve, Drakensberg**

The staff compound was continued to be renovated by crack filling of the walls. I later went to fetch a game guard on sick leave and take him along to a South African Police station in order to lay a charge against the person who had injured him. I also bought some nuts and springwashers for gate attachment.

9 August   **Monks Cowl Nature Reserve, Drakensberg**

I did an asset assessment of the station's saddles and other riding gear, including older items and those which had to be written off. I later checked on structural progress at the footpath, where shortcut routes taken by hikers or backpackers were stabilized using poles laid horizontally and perpendicularly to the slope. I also transported more poles to the path construction site.

10 August   **Monks Cowl Nature Reserve, Drakensberg**

I, assisted by a game guard, did intensive fence assessment on foot in order to obtain accurate estimates of the replacement cost along the boundary, from nearby Injasuti camp to Matterhorn.

11 August   **Monks Cowl Nature Reserve, Drakensberg**

I did intensive fence assessment from Steilberg at Barry's Grave to Culfargie. We could still find the land surveyor's pegs/beacons wherever there was no fence or parts thereof. I also verified Culfargie's assets in terms of saddles and smaller riding gear items.

12 August   **Monks Cowl Nature Reserve, Drakensberg**

Labour was organized for painting and alien plant control jobs. I subsequently took a horse with other equipment through to Culfargie. Upon returning to base, I did a vehicle inspection and checked the engine liquids. I then left for Giants Castle in order to do a resumed oribi count.

13 August   **Giants Castle Nature Reserve, Drakensberg**

I did my part of the oribi count at The Needle's foot near Giants Castle (peak). I recorded some blesbuck and eland. Later, after the count, we abseiled down a cliff in order to inspect a black eagle's nest. A chick was heard but not seen due to the difficult rock face.

14 August   **Monks Cowl Nature Reserve, Drakensberg**

Labour was organized for a range of jobs from crack filling, plastering and painting to alien plant control. I then repaired a broken signboard. Subsequently, I affixed Natal Parks Board decals to the doors of our official vehicle fleet and also checked the engine liquids. I started mapping and digitizing the fence assessment results achieved in the field. I also continued recording the impact of arson and other fires on the main map. I shared my list and suggestions of indigenous trees suitable for gardens in the region with a neighbouring land owner. I later went through a couple of aerial photographs in order to map out the footpath system around Sterkspruit. I consolidated the equestrian (saddlery) asset inventories of Monks Cowl and Culfargie stations. We combated a wildfire at Wonder Valley – it impacted upon the eastern bank of Spitsberg Stream and the adjacent valley (overtime: 1700-0000 hours).

15 August   **Monks Cowl Nature Reserve, Drakensberg**

After labour was organized for alien plant control, footpath construction and painting of the staff compound, I resumed my office analysis of the boundary fence in need of replacement. I then finalized the overall saddlery inventory. I also mapped the impact of the previous night's fire. Though relatively small locally, it was quite considerable outside the reserve. Two eland were seen in the region of Wonder Valley. After investigation of the affected area, the game guards patrolled the boundary to Injasuti camp in order to locate more surveyor pegs for complete fence assessment. I assisted with the transportation of vehicles to Spioenkop reserve for servicing, and made suggestions as to the items serviceable.

16 August   **Monks Cowl Nature Reserve, Drakensberg**

I continued my assessment of the boundary fence administratively (digitally). I analyzed photographic and other material, such as maps, in order to develop the footpath zone for day visitors. Previous maps of fire and other aspects of the reserve were interpreted, for instance for the period 1972-1977. Alien plant control involved Lantana or tick berry, American bramble, cotoneaster, bugweed, black wattle, and cluster pine. I also transferred Natal Parks Board emblems onto the remaining door panels of the station's vehicle fleet.

17 August   **Monks Cowl Nature Reserve, Drakensberg**

I extended the fence assessment on horseback from Culfargie's gate to Sikhaleni Senyoka at the boundary with Cathedral Peak Nature Reserve further north, but the fence was found to be non-existent. Afterwards, I finished some administrative tasks at Culfargie. Back at Monks Cowl station, I finalized my fence assessment and filed a comprehensive report. Fire impact analysis was resumed towards the end of the day.

18 August   **Monks Cowl Nature Reserve, Drakensberg**

I continued my fire impact analysis of the previous day by means of mapping. I then erected a signboard near the office and started working on more wooden signs using a router and a sander. I also assisted the officer-in-charge in laying a charge at the Winterton police station against poacher/s for illegal possession of game products. This involved statement giving, identifying confiscated items, and providing relevant information in connection with the crime committed. The items confiscated included a dead caracal, two dead duiker, baboon remains as well as slaughtering tools. Specific sections of the law violated and impending penalties were noted.

19 August   **Monks Cowl Nature Reserve, Drakensberg**

Building improvements at the staff compound continued and a door was hung, among other things. More campsite/office signboards were prepared by stencilling. Alien plant control also continued. I fixed the station's two-way radio (VHF transceiver) by securing the aerial and reattaching its cable.

20 August   **Monks Cowl Nature Reserve, Drakensberg**

I checked a few inventory items, such as radios, handcuffs, saddlery, and safes. I then assisted the visiting officials during their annual audit of the station's assets. After the audit had been completed, I fixed a vehicle. I subsequently hanged the door of a rondavel (hut) in the staff compound. I also supervised the painting and other interior renovations. I later completed the routing and sanding of two campsite/office signboards at the workshop.

21 August Monks Cowl Nature Reserve, Drakensberg

I continued with signboard making. I also facilitated the advice given by a professional paint consultant at the staff compound concerning the best primer, undercoat and topcoat to use on walls, doors, frames, and roofs. I later verified the asset inventory of the station's camping gear (incl. tents) and took stock of staff uniforms.

22 August Monks Cowl Nature Reserve, Drakensberg

Infestation of alien plants was investigated during a patrol on the northern and southern banks of Sterkspruit. I subsequently mapped and analyzed the findings which I had digitized. I continued with the signboards by filling with paint the letters that were grooved out by routing. I later inspected the staff compound and vehicle fleet. I also checked the chlorination levels of the water reservoir.

23 August Monks Cowl Nature Reserve, Drakensberg

Signboard making and painting continued. An estimate for a section of the boundary fence was added to the previous cost assessment. I phoned the state weather department to discuss the intricacies of the reserve's anemometer and totalizer responsible for counting the amount (kilometres) of wind passing through a locality (wind speed and wind prevalence). I also phoned automotive places concerning vehicle repairs, etc. I later fixed a toilet's flush handle/cistern float. I finished the day by checking an official vehicle's emergency gear.

24 August Monks Cowl Nature Reserve, Drakensberg

I checked the measurements for foundation casting of a new carport. I also verified that the weather recordings were being taken correctly for wind direction, total wind passage, minimum/maximum temperatures, wet/dry factor, hygrometric percentage (humidity), and cloud coverage (clear/overcast). Meanwhile, I had continued with signboard making.

25 August Monks Cowl Nature Reserve, Drakensberg

We combated three arson fires of which all three were started/flared up again in the same locality. I also checked on the weather recordings. I ended my duties for the day by inspecting an official vehicle's fluids and cleanliness.

26 August Monks Cowl Nature Reserve, Drakensberg

Alien plant control continued. I also went further with making signboards by sanding the rough edges left by the router before the letters were filled with paint. We later fought another arson fire at Culfargie (overtime: 1630-1930 hours).

27 August Monks Cowl Nature Reserve, Drakensberg

I organized labour for alien plant control. I then checked and filled a vehicle before picking up the game guards and labourers from Monks Cowl and Culfargie compounds to do shopping at Bergville and Estcourt. Other official duties were also taken care of.

28 August Monks Cowl Nature Reserve, Drakensberg

I went out and took the labourers to where the reserve had its boundary with some farmland in order to clear the fence line of wattle encroachment. This was as result of a mutual agreement between the farmer and us.

29 August Monks Cowl Nature Reserve, Drakensberg

I took the labourers to further the fence clearing that they had started on, and fetched them afterwards. Meanwhile, signboard making continued in terms of sanding and tidying the letterwork before applying knotting (sealer) to the wood. I also map-recorded the progress made on the front of alien plant control.

30 August Monks Cowl Nature Reserve, Drakensberg

I dropped the fence-clearing labour team and fetched them again. Signboard making involved varnishing of the wood. I also did a vehicle inspection. We cleaned the carburettor of a generator/compressor. I later phoned Sterling Saddles in connection with fitting sweat flaps to the new range of 'sergeant saddles.' I finalized the digitizing of the alien plant control undertaken. Priorities for block burning were also set up as the impact of the recent arson/accidental fires was assessed.

31 August Monks Cowl Nature Reserve, Drakensberg

I inspected the station's vehicles, filled them and fixed a bonnet release system. We also assessed the water supply system. The reservoir's outflow pipe seemed too small (35mm through) and an upgrade to a pipe of 50mm in diameter was needed. I later checked the sundries purchased against an invoice.

SEPTEMBER

1 September   Monks Cowl Nature Reserve, Drakensberg

The making of signboards progressed as letter filling with paint and applying of knotting continued. Horizontal edges of the signs were sanded at 45° angles from each side to form right angle vertices at the top and bottom in order to maximize vertical water runoff and reduce precipitation damage to signs. We later measured the distance for replacing the water reservoir's pipe with one having a larger circumference, as had previously been decided. I checked on the animal counts made by the game guards and recorded the baboons that were seen.

2 September   Monks Cowl Nature Reserve, Drakensberg

Labour was organized for breaking down an old vehicle inspection ramp. Additional road construction in terms of surfacing with fragmented quarry was carried out as well as the building of a carport. Signboard making also continued by painting of the lettering and varnishing of the wood. Towards the end of the day, I assisted at the entrance with admission payments.

3 September   Monks Cowl Nature Reserve, Drakensberg

We started block burning on the north-western side of Sterkspruit, but it died out due to showers. Signboard making continued in terms of painting. I finished the day's work by checking on the game guards' animal counts (common reedbuck were seen).

4 September   Monks Cowl Nature Reserve, Drakensberg

I supervised construction of the carport, but had to leave for Injasuti camp to do an asset assessment of the inspection quarters and building infrastructure. Thereafter, I went to Culfargie and issued the game guards with saddles and other riding gear for which they had to sign receipt.

5 September   Monks Cowl Nature Reserve, Drakensberg

I went to Pietermaritzburg for official business at Queen Elizabeth Park (head office). It was of paramount importance to acquire the deeds of properties on whose land a trail had been planned adjacent to an area owned by Natal Parks Board. I then obtained the official fire/weather recording forms from our regional scientist. I subsequently delivered a Monks Cowl vehicle to the Midmar workshop for repair of its differential. I also handed in our old .303 rifles. Back at base, I took two game guards to Culfargie to assist there in watching over impounded cattle.

6 September	Monks Cowl Nature Reserve, Drakensberg

I phoned a few people in connection with official events. I also reported theft of a signboard to the South African Police at Winterton. We resumed burning the north-western side of Sterkspruit so that the design of existing footpath routes could be easily evaluated and any additional layout properly planned. I later recorded the weather conditions and other data by filling out an official fire report form.

7 September	Monks Cowl Nature Reserve, Drakensberg

I furthered the mapping and digitizing of block burning and runaway/arson fires to date. I added to the botanical list of tree tags for the campsite and office gardens. I later issued saddlery to the game guards on base by having each of them sign receipt for their allocated gear. I ended the day's duties by deworming the horses and treating some bite (blowfly) wounds with antiseptic.

8 September	Monks Cowl Nature Reserve, Drakensberg

I took care of several administrative tasks relating to the time book/labour register, saddlery inventory, and fire mapping/priority recording form. I subsequently taught a member of staff how to read the weather instruments and record the data obtained. We later went out on horseback to patrol Matterhorn, where it was noticed that fire would hold no threat to Wonder Valley. Seven eland were seen in the area.

9 September	Monks Cowl Nature Reserve, Drakensberg

Labour was organized to clear the build-up of bat droppings within a roof, digging a soak pit for a shower at the office complex, cutting fence droppers out of wattle, and continuing road construction using quarry fragments. An alien plant control team was also set up at Culfargie. I went out on horseback patrol to the lookout via The Sphinx contour path to draw up a tourist map. This was aided by aerial photographs and done from an elevated vantage point, looking down on the footpaths below. Whilst up there, we also checked upon the repeater house. Five eland were seen on the plateau and three baboons in the vicinity. Previous problems of cattle entering this area of the reserve were discussed with the game guards. We identified three pine trees at Cathkin Peak for removal by the alien plant control team.

10 September	Monks Cowl Nature Reserve, Drakensberg

Labour was organized as per the previous day. The soak pit for the shower was completed. I assisted inbound calls by dispensing general advice about the reserve and its amenities (walks) and followed up with campsite reservations. I then tested the reservoir's water. The chlorine concentration was found to be 0.6ppm, although chlorine levels of 0.7ppm were required to ensure a low bacterial count safe enough for drinking. I later phoned the Midmar workshop to arrange wheel balancing/alignment as well as other automotive repairs. I also compiled a duty list for going to Pietermaritzburg. I completed the time book, labour register, and patrol register for the day and checked on the game guards' game counts. I filled a vehicle with sufficient fuel and topped up the oil.

11 September   **Monks Cowl Nature Reserve, Drakensberg**

I organized labour for alien plant control, in particular the killing of stands of American bramble by spraying the leaves with 2% Garlon in diesel. Staff compound renovations were resumed by preparing the walls with steel brushes for the application of primer, undercoat, and topcoat. In places, plastering of badly affected areas had to be done. I also assisted at the office complex as previously.

12 September   **Monks Cowl Nature Reserve, Drakensberg**

Labour was organized as for the previous day's jobs. In addition, the labourers retrieved the footpath compacters from the bush. Game guards were sent on horseback up to the plateau via The Sphinx in order to fell exotic pines, approximately five of them, at the foot of Cathkin Peak. I settled and finalized the saddlery issued to Culfargie game guards. I also checked and mapped/digitized the alien plant control done at Culfargie.

13 September   **Monks Cowl Nature Reserve, Drakensberg**

Signboard making resumed as more coats of varnish were applied and in higher concentrations. I subsequently checked on the renovations at the staff compound where preparation for the painting of more walls, beams and asbestos roofs took place. I checked the vehicle fleet and fixed a pair of windscreen-washing spray jets. I also assessed the condition of batteries with a battery tester (hydrometer). I then phoned Sterling Saddles to inquire about previous orders and adding to the existing order of saddlery. I later submitted a written report (compiled by the officer-in-charge) on Natal Parks Board's handling of the fence-clearing operation, as mutually agreed between us and the owner of the bordering farmland, in order to present our case against unsubstantiated accusations made by this person.

14 September   **Monks Cowl Nature Reserve, Drakensberg**

I checked on the staff compound renovations – roofs should have been scrubbed more and beams sanded more prior to painting. I also noticed emerging rust on an old water pipe near the tap, where it had to be replaced with a PVC section. I later assessed the footpath system at ground level by walking the different routes to verify and detail/digitize the photographical chart (bird's eye view). I increased the scale from 1:5000 to 1:10 000, making it coarser. Lastly, I investigated the edges of a bridge which had started to give way under heavy vehicle impact made worse by persistent rain.

15 September Monks Cowl Nature Reserve, Drakensberg

I furthered the ground-level assessment of the footpath system (Sterkspruit zone) by adding detail to the digitized map in order to make available to visitors of the reserve a complete yet simple map of the immediate area. This was eventually to be enhanced by the graphics division at Head Office. I later checked on the bridge repairs as well as on signboard varnishing. I also looked up fire management regulations as stipulated in the Forestry Act.

16 September Monks Cowl Nature Reserve, Drakensberg

I resumed digitizing the progress made in alien plant control and mapped the additional area which was cleared at Culfargie. I also filled out the official fire recording forms for which I had consulted the weather reports of the dates concerned. The game guards and I went out on horseback to the owner of the neighbouring property who had a tainted opinion of the reserve's dealings with him (same person as stated on the 13th). This was done to assess the amount of invasive fence-line vegetation cleared by Natal Parks Board compared to that by the adjacent landowner. I found that the eradication effort on our side had been of greater extent and that the relentless allegations would be overturned. I later explained the thatching requirements to the staff so as to enable them to supervise the rural women who harvest the thatch grass. I also checked on the game guards' patrol book. I then assessed the firebreaks which had been burnt.

17 September Monks Cowl Nature Reserve, Drakensberg

I went to Pietermaritzburg to do official business on behalf of the reserve. I also dealt with Natal Provincial Administration files. A compressor engine was handed in for servicing. General goods for maintenance were bought from Makro and FT Building Supplies. A front reflector for one of our vehicles was also bought. I changed the size of protective clothing for safe chainsaw operation previously purchased. The chainsaw's chain had to be sharpened too. Paint was collected to complete a transaction. Arms accessories were also collected. Bird netting was collected and a waistcloth ordered. Lastly, I purchased some stays and putty for windows after having dropped off a game guard at Albert Falls Nature Reserve for training.

18 September Monks Cowl Nature Reserve, Drakensberg

I filed the fire report of the previous day and mapped/digitized the extent of the burn. I subsequently checked on the renovations at the staff compound. I also checked and filled the vehicles. I then left to buy some maintenance tools at NLK Winterton and transported logs for firewood. I went through the window handles, stays and bolts at the office complex in preparation of new fittings.

19 September Monks Cowl Nature Reserve, Drakensberg

Official Day Off (spent at Ballito, Dolphin/North Coast).

20 September Monks Cowl Nature Reserve, Drakensberg

Official Day Off (spent at Ballito, Dolphin/North Coast).

21 September Monks Cowl Nature Reserve, Drakensberg

Official Day Off (spent at Ballito, Dolphin/North Coast).

22 September Monks Cowl Nature Reserve, Drakensberg

Official Day Off (spent at Ballito, Dolphin/North Coast).

23 September Monks Cowl Nature Reserve, Drakensberg

Official Day Off (spent at Ballito, Dolphin/North Coast).

24 September Monks Cowl Nature Reserve, Drakensberg

Official Day Off (spent at Ballito, Dolphin/North Coast).

25 September Monks Cowl Nature Reserve, Drakensberg

Official Day Off (spent at Ballito, Dolphin/North Coast).

26 September Monks Cowl Nature Reserve, Drakensberg

Official Day Off (spent at Ballito, Dolphin/North Coast).

27 September Monks Cowl Nature Reserve, Drakensberg

Official Day Off (spent at Ballito, Dolphin/North Coast).

28 September   Monks Cowl Nature Reserve, Drakensberg

Official Day Off (spent at Ballito, Dolphin/North Coast).

29 September   Monks Cowl Nature Reserve, Drakensberg

Official Day Off (spent at Ballito, Dolphin/North Coast).

30 September   Monks Cowl Nature Reserve, Drakensberg

Official Day Off (spent at Ballito, Dolphin/North Coast).

Official days off for July: 19-20 Sept
Official days off for Aug: 21-25 Sept
Official days off for Sept: 26-30 Sept

OFFICER IN CHARGE
MONKS COWL
Pvt. Bag X2 WINTERTON
3340 S. AFRICA

OCTOBER

1 October Sodwana Bay Marine Reserve, North Coast

Briefings were given to me as to how the entire region operates, especially the local reserve station. I contributed to beach patrolling and basic law enforcement by means of extension methods rather than prosecution. Illegal firewood collectors were confronted. Skippers without lifejackets were addressed. Fishing, swimming, scuba-diving, boating and driving zones (below high water mark) were pointed out by the officer-in-charge. I started reading through the management plan of the reserve, which detailed the range of the operational zone in terms of daily responsibilities. Brief history and reasons for the reserve's establishment were also taken note of. Diving and other restrictions were discussed. Speeding, waterskiing, motorbiking (incl. triking) and dune riding were strictly prohibited. Fishing regulations were briefly outlined to me – no bait was to be collected from the reserve.

2 October Sodwana Bay Marine Reserve, North Coast

I persisted in going through the management file, broadening my understanding of the turtle survey/monitoring programme in terms of species occurrence, distribution, migration, nesting paths, tagging, etc. I also maintained basic law enforcement on the beach – official presence. We transported brandering and slate tiles, as well as other items, for the construction of a watchtower at the beach and observed its erection by the building team. The importance of ideal sea conditions was noted for different activities, for example, when to go boating (low swell/not choppy), driving (low tide), fishing (high tide), or swimming (rising tide).

3 October Sodwana Bay Marine Reserve, North Coast

We drove past a vast number of campsites in order to assess which ones were occupied and which ones not. At the same time, any misconduct was taken note of, such as the use of neighbouring sites for parking. Also, the tidiness of ablutions was checked upon. I later transferred data onto the reservation chart in order to follow up new bookings and law enforcements.

4 October Sodwana Bay Marine Reserve, North Coast

I was shown the location of the reserve's radio repeater and water reservoir. We patrolled the beachfront on foot in order to conduct basic law enforcement, reiterating to people on the rocks that no bait or shells ought to be collected. A considerable number of ski-boats were checked so as to ensure that their basic emergency equipment was intact and complete. Action was taken against transgressors – notices were issued via the state department of transport. A written warning was issued to the diver who brought coral and shells out from the protected reefs. I maintained official beach presence in order to be available to render any extension in terms of advice, etc. Before the end of the day, I was briefed on yet more regulations and operational procedures.

5 October   **Sodwana Bay Marine Reserve, North Coast**

I patrolled a stretch of intertidal rocks on the beach with the same objectives as the day before. I also made sure that the boat launching area remained clear of unauthorized vehicles and people. Beach presence was asserted for extension purposes, giving advice where necessary.

6 October   **Sodwana Bay Marine Reserve, North Coast**

Beach presence was upheld for extension reasons and general law enforcement. I studied the obligations of skippers concerning onboard safety equipment for the different types of seaworthy vessels. I noted the light system required for vessels in terms of port and starboard. I also studied the fishing regulations, such as bag limits and closed season, depending on the species of fish.

7 October   **Sodwana Bay Marine Reserve, North Coast**

Additional tent campsites for the Blueprint Diving concession were created by aesthetically clearing some bush and levelling the ground. A way through the thickets was also cleared so as to set up a footpath to the ablution block. We checked out a power failure problem – wires leading from the main switchbox weren't connected – and reported it to a certified electrician.

8 October   **Sodwana Bay Marine Reserve, North Coast**

The team and I finished the work in the SLC camping area by means of clearing, levelling and footpath making as per the previous day. Official beach presence was later resumed for extension services to the public.

9 October   **Sodwana Bay Marine Reserve, North Coast**

We did a shoreline patrol by four-wheel drive vehicle from base to Mabibi (northwardly) and from base halfway to Red Sands (southwardly). Location names were studied and the turtle survey operations noted. Beach presence for public extension was also asserted. Law enforcement entailed the confiscation of mussels illegally removed from a rocky shore. The name and address of the perpetrator were recorded, though the offence was mitigated by an educational approach.

10 October   **Sodwana Bay Marine Reserve, North Coast**

Beach presence was resumed for policing and extension purposes. I briefed an aspiring student about the Nature Conservation career. I also advised others on beach regulations, weather reportings, and the like.

11 October   **Sodwana Bay Marine Reserve, North Coast**

Our beach presence was maintained during boat launching operations in order to ensure that all applicable regulations were adhered to. Law enforcement on the beach prevailed against irresponsible skippers who neither used lifejackets themselves nor advised their crew or passengers to comply. As a result, their particulars were taken down for prosecution. I also saw to it that the launching slip remained clear of obstructions.

12 October   **Sodwana Bay Marine Reserve, North Coast**

I maintained beach presence for extension purposes. General law enforcement concerning drunkenness and similar misconduct in public places was also conducted. I followed a new trail from the car park to the beach and noticed some surface erosion. I also ensured the clearing of the slipway for safe boating.

13 October   **Sodwana Bay Marine Reserve, North Coast**

Beach presence for public extension was maintained. Reasons for protecting the reefs (Quarter Mile, Two Mile, Five Mile, Seven Mile, and Nine Mile Reef) against fishing and harvesting – imposed by conservation regulations – were explained as different views arose around this topic. As usual, I kept the boat slipway clear. I assisted in towing a broken-down vehicle free from the sand and encroaching waves in order to get it to safety. I later assisted at the entrance complex.

14 October   **Sodwana Bay Marine Reserve, North Coast**

I patrolled the beach and asserted an official presence. I also assisted with corduroy track construction (using bluegum pole-mats) so as to make erosion-free, easy access possible to the beach. We winched out some visitors' vehicles that got stuck due to bulldozing works. Our vehicle's chassis was pressure-sprayed with detergent, followed by an oil-and-diesel mixture to minimize rust formation on surfaces prone to become salt-laden. I checked the connections on the exhaust system and found some missing springwashers from the manifold attachments.

15 October   Sodwana Bay Marine Reserve, North Coast

I undertook the daily beach patrol and maintained an official presence. I also ensured that the slipway remained safe for boat launching (outbound traffic) and beaching (inbound traffic). We later transported some goods. I assisted in native gardening using large-leaved dragon trees. We then visited the SLC camping ground. General regulations were enforced throughout the day.

16 October   Sodwana Bay Marine Reserve, North Coast

I patrolled the beach and remained present to make sure that safe boating operations took place. I also prevented people from accessing the area where the river mouth was going to be opened by a bulldozer, because tidal forces had rendered it impassable. Reasons for flattening the sand were given when asked by onlookers. We later unpacked the reserve's official diving and boating equipment so as to be ready for the next day's asset assessment. Lastly, netting for the turtle monitoring enclosures was washed.

17 October   Sodwana Bay Marine Reserve, North Coast

A routine beach patrol was undertaken and an official presence asserted. We rerolled the Land Rover's winch cable in order to neaten and straighten out any kinks, thereby preventing premature wear. We then went to the sawmill to buy the necessary timber and poles for making the annual turtle survey signs/markers. Back at base, I filed some documentation. We later checked on the building progress of one of the new staff houses. General law enforcement was carried out during the course of the day, where needed.

18 October   Sodwana Bay Marine Reserve, North Coast

Beach presence, beach patrol, general law enforcement, and slipway supervision were at the order of the day. Public extension was given where necessary. Boat launching times were discussed with the South African Police and compulsory cut-off times implemented. I had started to make the signboards/distance markers for the turtle survey by painting the surfaces in preparation of the numbers to follow in contrasting colour. I also assisted in corduroy road making and sand build-up at two off-road beach access sites by bulldozer.

19 October   Sodwana Bay Marine Reserve, North Coast

The beach was patrolled and an official presence maintained whilst the availability of the boat slipway was ensured. I also checked the boats' safety gear and searched the BC (buoyancy compensator) pockets of incoming divers for taken reef life (e.g. hard coral). No major violations occurred, and as such, no criminal prosecutions were made. I subsequently completed the signboards by painting on the numbers and nailing the boards onto single poles. We then visited the turtle survey site for the breeding season's enclosure (field hatchery) to be later erected. Meanwhile, general law enforcement and extension work was done.

20 October   **Sodwana Bay Marine Reserve, North Coast**

I did a regular patrol of the beach and maintained an official presence. A written warning was given to an offender who spear-fished for reef fish (galjoen). I attached a signboard onto poles for public display of the beach regulations. We also confiscated the angling gear that was used to catch undersized gamefish.

21 October   **Sodwana Bay Marine Reserve, North Coast**

Official Day Off (spent at Nongoma, KwaZulu-Natal Midlands).

22 October   **Sodwana Bay Marine Reserve, North Coast**

Official Day Off (spent at Nongoma, KwaZulu-Natal Midlands).

23 October   **Sodwana Bay Marine Reserve, North Coast**

Official Day Off (spent at Nongoma, KwaZulu-Natal Midlands).

24 October   **Sodwana Bay Marine Reserve, North Coast**

Official Day Off (spent at Nongoma, KwaZulu-Natal Midlands).

25 October   **Sodwana Bay Marine Reserve, North Coast**

Official Day Off (spent at Nongoma, KwaZulu-Natal Midlands).

26 October   **Sodwana Bay Marine Reserve, North Coast**

Beach patrol and general law enforcement were conducted. I assisted at the South Beach guard post in order to prevent divers, whose vehicles were not all wheel drive, from passing. I gave explanation to this being the first day of the regulation's implementation. Two-wheel-drive vehicles must remain in the parking lot, as offloading of diving gear closer to the beach was not allowed anymore. Beach access permits (KwaZulu) were issued and other passes checked in this regard. A newly-repaired and equipped patrol boat (Natal Parks Board) arrived and was given a check-over.

27 October   Sodwana Bay Marine Reserve, North Coast

The beach was patrolled and general laws enforced. Fishermen's bait was checked for sea-lice, which were illegal to have, but fortunately only sardines purchased from the bait shop were found. I also assisted at the guard post, stopping two-wheel-drive vehicles to go beyond the car park and offload dive equipment. Two offenders were given verbal warnings which could have led to them facing removal from the camp and reserve. We then planted the signposts with distance numbers for the turtle survey along the local coastline. I recorded one turtle coming ashore. The anglers' catch-of-the-day display was also visited. I finished the day's duties by doing some office administration.

28 October   Sodwana Bay Marine Reserve, North Coast

I performed the normal beach patrol. I also did a turtle survey patrol south of base – two nests were recorded: one of a loggerhead and one of a leatherback sea turtle. Back at base, I assisted in the general servicing of our Land Rover patrol vehicle with regard to new differential oil and gasket, gearbox oil, engine oil, clutch oil/brake fluid, air filter, and sparkplugs. The chassis was pressure washed and sprayed with oil to prevent rust.

29 October   Sodwana Bay Marine Reserve, North Coast

A beach patrol was done and an official presence maintained. Public extension regarding tides, boat operational zones and rural relations was provided. We carried out a turtle survey patrol northwards. Several loggerhead nests as well as a leatherback nest were seen. Some nesting paths were cancelled with an 'S' in the sand. I resumed reading through the management plan on turtle surveying and monitoring operations in terms of turtle tagging (different techniques for loggerheads and leatherbacks) and turtle clipping (juveniles). Our vehicle was washed as done daily. I also studied the reserve's most prevalent fish species.

30 October   Sodwana Bay Marine Reserve, North Coast

I conducted a beach patrol and asserted my presence for public extension, etc. We assisted in towing compromised vehicles and boats on trailers out of the soft sand due to neap tide which had caused a weak surface tension and hence loss of traction. We were also called out to sea to assist an injured seaman from a Cape Town trawler to board our vessel (courtesy of another skipper) for treatment on land. The MRI (Medical Rescue International) stabilized a bleeding fracture in the person's arm and ambulanced him to hospital.

31 October   Sodwana Bay Marine Reserve, North Coast

I did an early beach patrol at 0500 hours and remained present to see the boats of OET (Oos Eastern Transvaal) fishing club go out on the annual deep-sea bonanza for billfish/gamefish. Slipway management/beach control was conducted by the OET organizers. Back at the office, the necessary administration was done. I analyzed the angling/spear-fishing statistics, digitized the number of divers, launches and dives of the diving concessions (commercial enterprises), and statistically totalled and analyzed the data for April/August/September. The level of utilization during peak seasons could be clearly shown and the associated pressure on the reserve inferred.

Leatherback turtle digging out nest with game guard looking on from behind.

NOVEMBER

1 November   **Sodwana Bay Marine Reserve, North Coast**

The beach was patrolled and an official presence asserted. We conducted an investigation into an alleged theft of plant material from the reserve. No charge was laid, as the stolen plants were not recovered. I furthered the data digitizing for October as per the previous day. Lastly, we prepared the station's boat for future launching, checking the safety gear, fuel and engine oil.

2 November   **Sodwana Bay Marine Reserve, North Coast**

I conducted a patrol of the beach and maintained an official presence. Law enforcement in connection with regulated spear-fishing was performed through educational extension. We went out on sea patrol with our revamped boat to test the new motors and newly-fit equipment. The gear cable attachments were adjusted in the control box and engines and the idling was set. Law enforcement at sea was also carried out – the alpha flag wasn't flown by some skippers when their divers were down for which they were sternly reprimanded. Back at shore, the saltwater was cleaned from our boat and vehicle and a check-up on the engines done. The Land Rover's battery terminals had to be replaced, as they caused dangerous acidic fumes. I went for a turtle survey patrol at night, but no turtles were found. Thus, none could be tagged or recorded.

3 November   **Sodwana Bay Marine Reserve, North Coast**

Beach presence was maintained after a patrol had been conducted. We then carried out a turtle survey patrol southwards all the way to Red Sands – two leatherback nests were spotted and recorded. Upon returning, I did a vehicle check-up and top-up of the necessary fluids. Later on, we created a seedbed for the new lawn of a staff house by spreading debris across the prepared surface. At night, a turtle survey patrol was done, extending considerably north from base and including a 5km-stretch southwards. However, no sightings were made.

4 November   **Sodwana Bay Marine Reserve, North Coast**

I conducted a patrol of the beach and remained present for general law enforcement. We subsequently went out to sea to collect species of red algae (Rhodophyta) from the reefs as a research expedition, during which I functioned as second skipper (top-man) for the diving crew. After our return to shore, the boat and engines were washed, rust-guarded and made ready (fuel/oil ratio) for the next excursion. I also checked the vehicle for the turtle survey. By nightfall a turtle survey patrol was underway. One loggerhead sighting eventuated, which allowed us to tag and record it, as well as for one leatherback.

5 November   **Sodwana Bay Marine Reserve, North Coast**

Official beach presence was asserted. We also attempted to identify the cause of our faulty vehicle by carrying out various troubleshooting checks. The non-start was found to be due to a defective fuel pump. We subsequently went out to sea with the same research objectives as the previous day – exploration of the shallow Two Mile Reef. Back at shore, we washed the boat, trailer and vehicle and took preventative care against rust. Diving equipment was also cleaned of the saltwater.

6 November   **Midmar Dam Nature Reserve, Midlands**

The chief conservator and I travelled to the Midmar automotive workshop in order to collect a vehicle for basic use at Sodwana Bay. Career opportunities and job requirements were discussed on the way. I also attended the meeting with Phinda Nyala Resource Reserve, discussing possibilities around block bookings for lake tours. Boat cruising for the entire estuary complex was also covered. After having arrived at Midmar, I was accommodated in a chalet at the waterfront.

7 November   **Sodwana Bay Marine Reserve, North Coast**

I drove the vehicle scheduled for beach patrol from Midmar to Sodwana. Upon reaching Sodwana, a truck's manifold and starter were loaded onto another vehicle made ready to depart for Empangeni very early the next morning. An engineering service was going to be met to test the overhauled truck engine. Lastly, we checked on the local purification plant.

8 November   **Sodwana Bay Marine Reserve, North Coast**

We went to Empangeni to fetch the truck engine after having been tested. A Leyland fuel pump and battery were also purchased. On the return journey we bought some oil from Total Hluhluwe. Back at base, the engine was offloaded using a pulley crane. At night I embarked on a turtle survey by vehicle. One loggerhead was sighted, but no nesting occurred due to tourist intrusion. An educating reprimand was given in this regard. Fortunately, the turtle could be tagged.

9 November   **Sodwana Bay Marine Reserve, North Coast**

I asserted an official presence on the beach before assisting in truck engine mounting as well as gearbox alignment and fitment (incl. propshaft, alternator, starter, radiator, etc). Back at the beachfront we removed the windows from the watchtower in order to make better viewing (free of sea spray) possible for beach control. Long after sunset, a turtle survey was conducted during which two loggerhead sightings were made. One turtle nested and was tagged and measured.

10 November   **Sodwana Bay Marine Reserve, North Coast**

As usual, I started the day with a beach patrol and remained present to give public extension where necessary. We then assembled the planks for a desk inside the control hut. A turtle survey patrol south of base was also carried out: several leatherback nests were sighted. One turtle was tagged and her measurements were taken after she had nested. At night we undertook another turtle survey south of base during which one leatherback nest was recorded. Informative extension was given and law enforcement applied to offenders parked above the high water mark between the dunes.

11 November   **Sodwana Bay Marine Reserve, North Coast**

I did a morning patrol of the beach and maintained an official presence for general law enforcement. I also supervised repairs of the corduroy pole track leading onto the beach. My assistance was subsequently required for fuel pump fitment at the workshop. Back at the beach, I represented the Board at the weigh-in of the Billfish 5000 fishing competition

12 November   **Sodwana Bay Marine Reserve, North Coast**

I prepared for and assisted and catered for a stakeholder delegation surrounding land negotiations near False Bay (rural district). Upon our return to base, we had to unpack everything.

13 November   **Sodwana Bay Marine Reserve, North Coast**

We carried out an extensive beach patrol and asserted an official presence. The coastline all the way to Leven Point was covered. Unexplained whale shark beachings were investigated. Signs of high parasite loads and haemorrhagic septicaemia hinted at the cause of the deaths.

14 November   **Sodwana Bay Marine Reserve, North Coast**

I conducted a beach patrol and remained present to clear the boat slipway and give public extension where required. We also fixed the ladder of the control tower on the beach. A signboard was mounted for display in connection with closed beaches (at night), effective from the fifteenth of November. I ended the day by reading up on the southern African sea fishes, specifically the whale shark.

15 November   **Sodwana Bay Marine Reserve, North Coast**

Beach patrol, official presence and public extension were at the order of the day. The officer-in-charge checked the seaworthiness of a new ski-boat belonging to a visitor and by special permission granted that it could go to sea without prior registration. I subsequently took the ticket books to the entrance gate and continued with general administration. I organized the squaredavel's beds to accommodate an influx of guests. Beaches were closed at night and signs posted due to the turtle nesting season which had commenced.

16 November   Sodwana Bay Marine Reserve, North Coast

I carried out a patrol on the beach and remained present for public extension. Regulations surrounding the wearing of lifejackets were explained and warnings given to the offending boat crew. Drivers of vehicles above the high water mark were also reprimanded. We later changed the wheels of the turtle translocation vehicle with broader ones for better capability on sand. Repairs of the southern guard post hut, which was sabotaged the previous night, were also arranged.

17 November   Sodwana Bay Marine Reserve, North Coast

Beach patrol, official presence, public extension and law enforcement were the duties set for the day. I checked the safety gear of dive boats and the BC (buoyancy compensator) pockets of divers returning to shore, but no offences were committed. Vehicles parked in restricted zones were issued with notices of warning. An issue around the display of a banner was resolved in terms of beach advertising not being allowed. Spear-fishing, illegal on Two Mile Reef, was also checked upon – no transgressions occurred.

18 November   Sodwana Bay Marine Reserve, North Coast

My default position was the beach – patrolling and remaining visible and vigilant – dispensing advice and enforcing the law where necessary. I issued written warnings to parking offenders. Offenders from boats were also warned for not wearing their lifejackets. After the last sabotaged signboards had been restored and the debris cleaned up at South Beach guard post, I took the permit money to the office to be put on safe deposit.

19 November   Sodwana Bay Marine Reserve, North Coast

I embarked on a beach patrol and maintained a presence for purposes of general law enforcement and extension. We started preparing our Land Rover for write-off and going to auction due to age. Fitted equipment, such as winch and towbar, was stripped. We then went to the sawmill nearby to consider suitable material on offer for constructing the turtle translocation enclosure. Treated and untreated wood depots were visited and an order placed.

20 November   **Sodwana Bay Marine Reserve, North Coast**

We went to Mbazwana Sawmill to fetch ordered poles and purchase more. Upon returning to base, we continued to strip the Land Rover (rollbar, tyres, battery, 2-way radio) and cleaned it out for transportation to the Midmar workshop. Onto a different task, I started familiarizing myself with the detail in seafaring regulations, based on the different size categories of vessels, so as to inform aspirant and bona fide boat owners of such stipulations when visiting Sodwana. Finally, our vehicle was loaded by ramp onto a flatbed truck.

21 November   **Sodwana Bay Marine Reserve, North Coast**

I conducted the usual beach patrol, etc. An insentient man with air (nitrogen) in the blood (decompression sickness or DCS), due to incorrect surfacing from a scuba dive, was attended to by MRI (Medical Rescue International). Later, at the workshop, I painted the roof-frame of the turtle tour vehicle with primer and gave it a final coating. I also riveted the front numberplate onto the vehicle's bumper. I then wrote a letter to a member of the public wanting to take his catamaran sailing vessel to sea. Persons were warned not to fish with nets and that they would be prosecuted on the next offence. I finished the day by assisting at the gate of the main office complex. I also digitized the ski-boat and fishing data.

22 November   **Sodwana Bay Marine Reserve, North Coast**

Patrolling, sustained presence, general law enforcement and extension were exerted on the beach. We surveyed the northern shoreline all the way to Mabibi – no turtle occurrences. After returning to base, I assisted the officer-in-charge to install the VHF (very high frequency) radio into our new vehicle (Land Cruiser). I also assisted at the chalet office.

23 November   **Sodwana Bay Marine Reserve, North Coast**

I maintained an official presence on the beach and performed general law enforcement. Intensive extension was given to members of the public. I also ensured that the boat slipway remained open for launching and beaching. I checked the BC (buoyancy compensator) pockets of returning divers for any reef life taken – two offences had been committed and stern warnings, even of police intervention, were given. Offenders claiming ignorance of lifejacket negligence were also warned. I forbade unregistered boats to go out to sea. I then assisted in a Natal Parks Board scuba dive and beaching operation.

24 November   **Sodwana Bay Marine Reserve, North Coast**

I ensured an official presence and control on the beach as well as public extension and general law enforcement. The 'rubber-duck' boats were checked for seaworthiness – a few violations of the regulations occurred. I again provided logistical support to the Natal Parks Board dive team.

25 November Sodwana Bay Marine Reserve, North Coast

Beach presence, public extension and general law enforcement were the standard duties of the day. A turtle survey patrol to Mabibi was carried out during which several sightings of loggerhead turtles were made, but no nesting occurred. Only one leatherback nest was noted. At night, another turtle patrol northwards was undertaken. One loggerhead was tagged, while several tracks and nests of the same species were recorded.

26 November Sodwana Bay Marine Reserve, North Coast

I maintained an official presence on the beach for purposes of general law enforcement. We then did a turtle patrol along the northern shore. One loggerhead was sighted and tracks and nests of both the species were come across. We also went to Manzengwenya to pick up the Mabibi permits from KwaZulu Forest Reserve.

27 November Sodwana Bay Marine Reserve, North Coast

I performed a beach patrol and remained present for general law enforcement and public extension. I cleared the windsurfers and swimmers from the boat slipway zone. We later looked for a suitable tree or plant to cut straight, durable sticks from for holding up number signs relating to turtle nests of the survey/translocation programme.

28 November Sodwana Bay Marine Reserve, North Coast

We carried out a beach patrol, turtle patrol, and general law enforcement. Approximately 200 sticks were cut for marking nests of the turtle survey. We subsequently visited the translocation site along the shore to finalize the specifications of the enclosure to be erected. Back at base, the vehicles were washed, checked and filled. After heading out again, the lingering smell of a decomposing whale shark was eradicated by burying the remaining head under the sand.

29 November Sodwana Bay Marine Reserve, North Coast

A beach patrol was conducted, general law and order maintained, and public extension given where lacking. I started erecting the turtle relocation enclosure by putting in the corner posts, straining posts, and inline carrier poles. One strand of wire had been pulled before we embarked on a turtle patrol. Upon returning, the vehicles were washed, etc.

30 November   **Sodwana Bay Marine Reserve, North Coast**

I continued work on the turtle enclosure by hanging sections of netting and burying the bottom part thereof in the sand. When returning to the recreational area to resume an authoritative presence, a beach patrol was conducted and general law enforcement carried out. Driving above the high water mark, as well as paragliding (slope soaring) and sand boarding, was addressed. I did an interpretive turtle tour for overseas guests while carrying out a shoreline survey at night. One leatherback was tagged and several nests of this species were seen.

Whale shark post-mortem by two marine biologists.

Turtle enclosure for hatchling monitoring.

DECEMBER

1 December   **Sodwana Bay Marine Reserve, North Coast**

A patrol was done along the beach where I remained present for public extension and general law enforcement. I finished erecting the turtle enclosure by burying the rest of the netting down at the bottom, and sewing all the sections together. Back at base, I briefed a student thoroughly on beach duty and related responsibilities. I also neatened the roof canvas of the turtle tour vehicle and equipped it with any outstanding emergency items. After sunset, I took a group of tourists out on a turtle survey. A loggerhead was seen which had already been tagged. Law enforcement was subsequently carried out, as overnighting and making of fire on the beach as well as unofficial driving at night beyond Jesser Point and Sodwana Bay were strictly prohibited.

2 December   **Sodwana Bay Marine Reserve, North Coast**

I maintained an official presence after having patrolled the beach. General law enforcement and public extension were also performed. I checked the spear-fishing licenses of competition entrants and the fish they shot. Only gamefish were allowed to be taken; no reef or bottom fish. Zero violations occurred in this regard. Meanwhile, our vehicle's cleaning, checking and filling up were done back at the workshop. I also made a small gate for the turtle enclosure, thereby finishing its construction entirely. Labour was assigned to the game guards for what remained left of the day.

3 December   **Sodwana Bay Marine Reserve, North Coast**

Beach work was done as per usual. Access to unwanted footpaths was blocked off with thorn bush to limit people to the boardwalks, thus eliminating dune erosion. I subsequently made the vehicles ready for the evening's turtle survey/tour. We then went to the sawmill to purchase poles and to Mbazwana Post Office for mail, etc. At nightfall, I went out to conduct a turtle tour. Unfortunately no turtles were sighted, but intensive extension was provided to the group nonetheless.

4 December   **Sodwana Bay Marine Reserve, North Coast**

Beach work was done as previously. Rails of poles were erected in order to cordon off a tourist area where no vehicles were allowed. I also assisted in taking a buoy out to sea, positioning and anchoring it to the seabed so as to demarcate the boating zone more clearly with relation to the beach.

5 December Sodwana Bay Marine Reserve, North Coast

Official Day Off (travelling back to hometown Pretoria).

6 December Sodwana Bay Marine Reserve, North Coast

Official Day Off (spent in hometown Pretoria).

7 December Sodwana Bay Marine Reserve, North Coast

Official Day Off (spent in hometown Pretoria).

8 December Sodwana Bay Marine Reserve, North Coast

Official Day Off (spent in hometown Pretoria).

9 December Sodwana Bay Marine Reserve, North Coast

Official Day Off (spent in hometown Pretoria).

10 December Sodwana Bay Marine Reserve, North Coast

Official Day Off (spent in hometown Pretoria).

11 December

Annual Leave (spent in hometown Pretoria).

12 December

Annual Leave (spent in hometown Pretoria).

13 December

Annual Leave (spent in hometown Pretoria).

14 December

Annual Leave (spent in hometown Pretoria).

15 December

Annual Leave (spent in hometown Pretoria).

16 December

Annual Leave (spent in hometown Pretoria).

17 December

Annual Leave (spent in hometown Pretoria).

18 December

Annual Leave (spent in hometown Pretoria).

19 December

Annual Leave (spent in hometown Pretoria).

20 December

Annual Leave (spent in hometown Pretoria).

21 December

Annual Leave (spent in hometown Pretoria).

22 December

Annual Leave (spent in hometown Pretoria).

23 December

Annual Leave (spent in hometown Pretoria).

24 December

Annual Leave (spent in hometown Pretoria).

25 December

Annual Leave (spent in hometown Pretoria).

26 December

Annual Leave (spent in hometown Pretoria).

27 December

Annual Leave (spent in hometown Pretoria).

28 December

Annual Leave (spent in hometown Pretoria).

29 December

Annual Leave (spent in hometown Pretoria).

30 December

Annual Leave (spent in hometown Pretoria).

31 December

Annual Leave (spent in hometown Pretoria).

[Handwritten annotations: signature dated 4/12/91; "Official days off / LEAVE : 5-10 Nov 1991 : 11-31 Dec 1991"]

Romans 16:27   *To the only wise God be glory forever through Jesus Christ! Amen.*

BIOGRAPHY

Victor Meyer has been involved in conserving nature since the late 1980s. He successfully completed several formal training courses, from a National Diploma in Nature Conservation (the focus of this book) to a Masters degree (cum laude) in the same field. His scientific research on termites in Kruger National Park, South Africa, has led to a PhD in ecological Entomology from the University of Pretoria in 2002, after being granted numerous sponsorships and scholarships. Dr Meyer has published internationally in peer-reviewed science journals. He now lives with his wife Debbie and daughter Nadine in Auckland, New Zealand, where he has been serving as Habitat editor for the Royal Forest and Bird Protection Society for close to six years. He is also a recipient (2007/08) of a research grant from the JS Watson Conservation Trust for collaborative work on terrestrial isopods (Crustacea), emanating from his discovery of a new species from South Africa in 1995.

Scientific Publishing House
offers
free of charge publication

of current academic research papers, Bachelor´s Theses, Master's Theses, Dissertations or Scientific Monographs

If you have written a thesis which satisfies high content as well as formal demands, and you are interested in a remunerated publication of your work, please send an e-mail with some initial information about yourself and your work to *info@vdm-publishing-house.com*.

Our editorial office will get in touch with you shortly.

VDM Publishing House Ltd.
Meldrum Court 17.
Beau Bassin
Mauritius
www.vdm-publishing-house.com

Made in the USA
Middletown, DE
13 May 2021